SS7 Basics

Third Edition

By

Lawrence Harte
Richard Dreher
David Bowler
Toni Beninger

Published By:
Althos Publishing
404 Wake Chapel Road
Fuquay-Varina, NC 27526 USA
Telephone: 1-800-227-9681
Fax: 1-919-557-2261
email: Success@Althos.com
web: www.Althos.com

Althos

International Standard Book Number: 0-9728053-7-0

Acknowledgements

We thank the many gifted people who gave their technical and emotional support for the creation of this book. In many cases, published sources were not available on this subject area. Experts from manufacturers, service providers, trade associations and other telecommunications related companies gave their personal precious time to help us and for this we sincerely thank and respect them.

Special thanks to the people who assisted with the production of this book including: Tom Pazderka (illustator), Jackie Gottlieb (graphics artist), and Bill Stevens (cover artist). Thanks to Karen Bunn (editor) at Althos who helped ensure this book was at the highest industry standard and that the book contained valuable and quality information.

About the Authors

Mr. Harte is the president of Althos, an expert information provider covering the communications industry. He has over 29 years of technology analysis, development, implementation, and business management experience. Mr. Harte has worked for leading companies including Ericsson/General Electric, Audiovox/Toshiba and Westinghouse and consulted for hundreds of other companies. Mr. Harte continually  researches, analyzes, and tests new communication technologies, applications, and services. He has authored over 30 books on telecommunications technologies on topics including Wireless Mobile, Data Communications, VoIP, Broadband, Prepaid Services, and Communications Billing. Mr. Harte's holds many degrees and certificates include an Executive MBA from Wake Forest University (1995) and a BSET from the University of the State of New York, (1990).

Richard C. Dreher, P.E. is an executive consultant to the telecommunications industry working for global telecommunication companies. Since 1983, Mr. Dreher has experience in digital communications, including long distance tandem switching, local number portability (LNP), local area network (LAN) design, SS7 engineering, cellular equipment product planning and technical sales, and Advanced Intelligent Network (AIN) and Service Creation product marketing. His  entrepreneurialism led to start-up experience with a PCS wireless carrier as their network systems technology director. Richard earned his BSEE from the University of Colorado, is a registered Professional Engineer (P.E.), a senior member and past executive officer of the Institute of Electrical and Electronics Engineers (IEEE), and a published author in trade journals, text books, and magazines.

 Mr. Bowler is an independent telecommunications training consultant. He has almost 20 years experience in designing and delivering training in the areas of wireless networks and related technologies, including CDMA, TDMA, GSM and 3G systems. He also has expertise in Wireless Local Loop and microwave radio systems and has designed and delivered a range of training courses on SS7 and other network signaling protocols. Mr. Bowler has worked for a number of telecommunications operators including Cable and Wireless and Mercury Communications and also for Wray Castle a telecommunications training company where he was responsible for the design of training programmes for delivery on a global basis. Mr. Bowler was educated in the United Kingdom and holds a series of specialized maritime electronic engineering certificates.

Toni Beninger is an interactive multimedia software specialist for the telecommunications industry. Ms. Beninger earned her B.Sc. degree from the University of Ottawa, and subsequently completed her education qualifications at Queen's University. She has had an extensive career in telecommunications documentation and training. Ms. Beninger was the senior technical training manager at Bell-northern Research Ltd.'s main laboratory in Ottawa. Ms. Beninger has been an executive at Sanctuary Woods, a multimedia software company. Ms. Beninger has extensive experience in SONET and Signaling System 7 (SS7).

Table of Contents

PREFACE

Signaling system 7 (SS7) is the standard communication system that has been used to control public telephone networks since the 1980s. In addition to the control of voice calls, SS7 technology now offers advanced intelligent network features and it has recently been updated to include broadband control capabilities, local number portability, and mobile communication services. SS7 networks are now interconnecting with and operating on Internet data networks. This book provides an introduction to SS7 systems and how to interconnect SS7 systems to and through other networks.

SS7 Basics, 3rd Edition provides a brief introduction to SS7, allowing the reader to gain an overall appreciation and understanding of its structure through the use of step-by-step procedures that describe the actions that occur in the network. It covers the reasons why SS7 exists and how it works along with recent developments. It is well suited for the technical and the non-technical readers who are involved in public telephone and voice over data networks and systems.

The chapters in this book are organized to help professionals in the telecom and communication related industries to get a well-rounded viewpoint of SS7 signaling systems and how recent industry changes have affected SS7. Chapters are divided into introduction overview, each SS7 layer, applications that use SS7, mobile communications, intelligent networks, SS7 and Internet Protocol, and international SS7 systems. The chapters may be read either consecutively or individually.

Chapter 1. SS7 Defined. Provides an introduction to SS7 technology and its system architecture. It covers the network elements and link types that connect them. Basic SS7 network functions such as message routing and information services are explained. A comparison between the SS7 protocol stack and the OSI 7 layer system is provided.

Chapter 2. Message Transfer Part, Level 1. Describes the physical layers used by SS7 systems. This includes analog and digital signaling links along with the typical data rates and characteristics.

Chapter 3. Message Transfer Part, Level 2. Explains the structures of the three basic types of packets LSSU, FISU, and MSU. This chapter provides fundamentals signaling on SS7 data links. Overviews of link setup and flow control procedures are provided.

Chapter 4. Message Transfer Part, Level 3. This chapter describes the management of message routing within SS7 networks. This detailed chapter explains the handing of signaling messages in SS7 network nodes. Signaling link identification, message routing, message distribution, and link management is described and analyzed. Link rerouting procedures and permissions are reviewed and an introduction to route sets and is provided.

Chapter 5. Signaling Connection Control Part (SCCP). This chapter provides an overview of how the SCCP identifies, labels, translates, and controls logical signaling channels in the SS7 system. It covers connection-oriented and connectionless services. The SCCP packet formats are explained and sample call flow sequences are provided.

Chapter 6. Integrated Service Digital Network User Part (ISDN-UP). This is an introduction to basic voice call processing. This chapter explains the setup of basic bearer services for voice and data communication. Sample call setup, connection management, and call release sequences are included. The ISDN-UP message format and CIC identification codes are described.

Chapter 7. Transaction Capabilities Applications Part (TCAP). A primer on how applications use SS7 signaling to provide non-circuit related information between applications. This chapter discusses how TCAP provides for advanced services such as 800 number, calling card, and Freephone service. The message formats used for TCAP messages are detailed along with sample TCAP message flows.

Chapter 8. Operation, Administration, and Maintenance Part (OMAP). Provides an overview of the management of SS7 systems including the management model, management information base (MIB), and application service elements (ASE.) SS7 network testing is described and message routing verification test (MRVT) and circuit validation test (CVT) sample steps are provided.

Chapter 9. SS7 Applications. Explains the fundamentals of providing applications through the use of SS7 signaling. Sample applications that are described include voice mail integration, network automatic call distribution (NCAD), virtual private networking (VPN), custom local area signaling services (CLASS), enhanced 800 number services, and calling card services.

Chapter 10. Mobile Applications Part (MAP). This chapter covers the use of SS7 signaling for mobile voice and messaging services. It describes how SS7 in used to satisfy the unique requirements of mobile devices. Covered is the integration of MAP in the SS7 protocol stack.

Chapter 11. Local Number Portability (LNP). Provides the fundamentals for number portability and the network elements that are involved in porting telephone numbers. Covers system portability, service portability, and geographic portability.

Chapter 12. Broadband ISDN User Part (B-ISUP). Provides the fundamentals of how SS7 has been enhanced to provide controls for broadband communication systems. A brief introduction to ATM technology and new SS7 broadband messages is provided.

Chapter 13. Intelligent Networks. Covers the use of triggers and intelligent applications to provide for advanced services. Covered is Service Creation Environment (SCE), Service Independent Building-blocks (SIBs) and how they are used in Intelligent Networks (IN), Customized Applications for Mobile Networks (CAMEL), and Wireless Intelligent Networks (WIN).

Chapter 14. SS7 and Internet Protocol (IP). Provides an introduction for transporting SS7 over IP and interfacing SS7 with other systems. This chapter covers the signaling transport (SIGTRAN) system and it's use of SCTP packet transmission. Explained are the adaptation layers M3UA, M2UA, M2PA, SUA, and IUA that are used to interconnect SS7 nodes via IP systems. Overviews of SIP, MGCP, MEGACO, and H.323 are included.

Chapter 15. International SS7. Identifies some of the key differences between SS7 systems used throughout the world. It describes different point code addressing systems, circuit identifiers, media transcoding requirements, and international signaling gateways.

A detailed glossary is included that can be used as a reference for key terms and acronyms associated with SS7 and other signaling system.

Chapter 1

SS7 Defined

Common Channel Signaling is a signaling method in which a signaling channel conveys, by means of labeled messages, signaling information relating to call setup, control, network management, and network maintenance. Examples of Common Channel Signaling systems are CCITT Signaling System No. 7 and various national versions such as ANSI SS7 as originated by the original Bell Communications Research (now named Telcordia Technologies) and AT&Ts original SS6 and SS7 standards.

The Need for SS7

Worldwide telephone networks are undergoing significant changes as methods of call processing and network management are altered to provide new services and to streamline operations. These changes are driven by user demand for enhanced services and the corresponding efforts of telephone operating companies to satisfy current and future needs. Enhanced services require bi-directional signaling capabilities, flexibility of call setup, and remote database access.

Earlier signaling systems lacked the sophistication required to deliver much more than plain old telephone service (POTS). These traditional systems use dial pulses and multi-frequency (MF) tones to transmit call and circuit-related information such as dialed digits and circuit busy/idle states.

The complexity of adding new functionality to traditional signaling systems meant that a new network signaling architecture was needed. SS7 was developed to satisfy the telephone operating companies' requirements for an improvement to existing signaling systems.

Basic SS7 Network Architecture

A telecommunications network consists of a number of switches and application processors interconnected by transmission circuits. The SS7 network exists within the telecommunications network and controls it. SS7 achieves this control by creating and transferring call processing, network management, and maintenance messages to the network's various components.

Figure 1.1 shows that a SS7 network has three distinct components: Service Switching Points (SSP), Signal Transfer Points (STP), and Service Control Points (SCP). These components may be generically referred to as "nodes" or "signaling points."

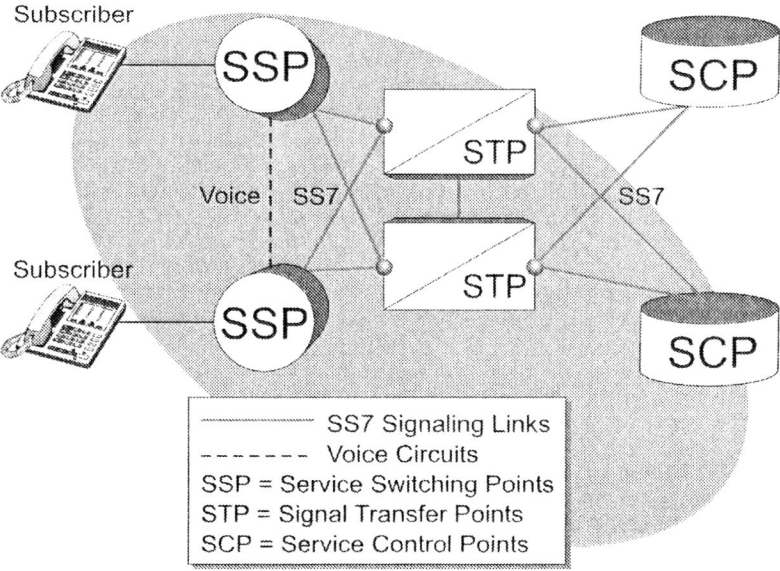

Figure 1.1. SS7 Signaling Network Architecture

SS7 Signaling Link Types

The lines that transfer messages in the SS7 system are called signaling links. Signaling links are logically organized by link type ("A" through "F") according to their use within the SS7 signaling network.

Figure 1.2 shows the relationship between the link names and the link location (type). Signaling links are logically organized by link type ("A" through "F") according to their use in the SS7 signaling network. The "A" (access) links connect the signaling end points (e.g., an SCP or SSP) to the STPs. Only messages originating from or destined to the signaling end point are transmitted on an "A" link. The "B" (bridge) links connect the STP to anoth-

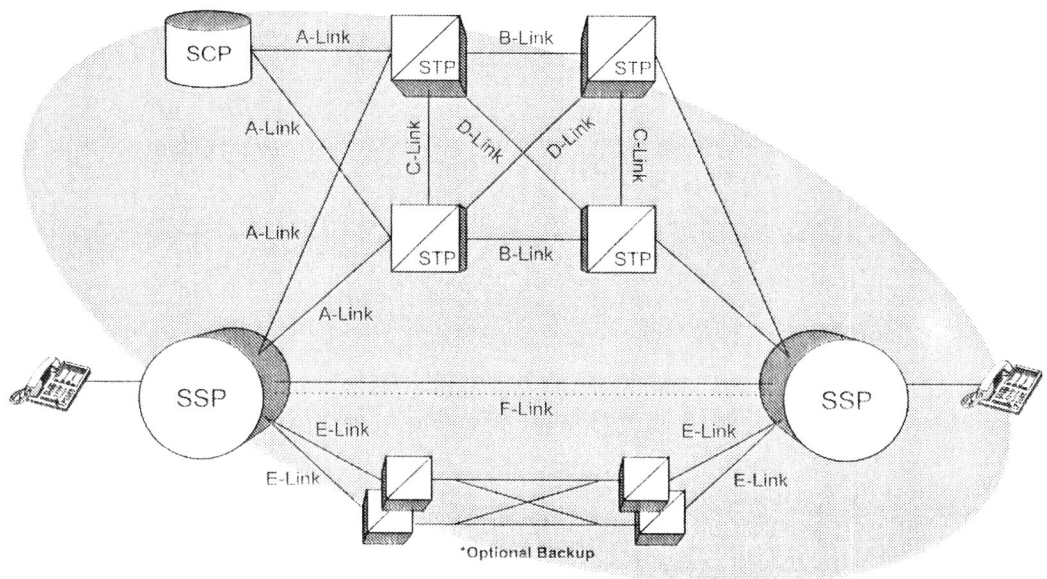

Figure 1.2. SS7 Signaling Link Types

er STP. The "C" (cross) link connects STPs performing identical functions into a mated pair. "D" (diagonal) links connect the secondary (e.g., local or regional) STP pair to a primary (e.g., inter-network gateway) STP pair in a quad-link configuration. "E" (extended) links connect the SSP to an alternate STP. An "F" (fully associated) link is connected between two signaling end points (i.e., SSPs and SCPs).

Access Link (A Link)

An "A" (access) link connects a signaling end point (e.g., an SCP or SSP) to an STP. Only messages originating from or destined to the signaling end point are transmitted on an "A" link.

Bridge Link (B Link)

A "B" (bridge) link connects an STP to another STP. Typically, a quad of "B" links interconnect peer (or primary) STPs (e.g., the STPs from one network to the STPs of another network). The distinction between a "B" link and a "D" link is rather arbitrary. For this reason, such links may be referred to as "B/D" links.

Cross Link (C Link)

A "C" (cross) link connects STPs performing identical functions into a mated pair. A "C" link is used only when an STP has no other route available to a destination signaling point due to link failure(s). Note that SCPs may also be deployed in pairs to improve reliability; unlike STPs, however, mated SCPs are not interconnected by signaling links.

Diagonal Link (D Link)

A "D" (diagonal) link connects a secondary (e.g., local or regional) STP pair to a primary (e.g., inter-network gateway) STP pair in a quad-link configuration. Secondary STPs within the same network are connected via a quad of "D" links. The distinction between a "B" link and a "D" link is rather arbitrary. For this reason, such links may be referred to as "B/D" links.

Extended Link (E Link)

An "E" (extended) link connects an SSP to an alternate STP. "E" links provide an alternate signaling path if an SSP's "home" STP cannot be reached via an "A" link. "E" links are not usually provisioned unless the benefit of a marginally higher degree of reliability justifies the added expense.

Fully Associated Link (F Link)

An "F" (fully associated) link connects two signaling end points (i.e., SSPs and SCPs). "F" links are not usually used in networks with STPs. In networks without STPs, "F" links directly connect signaling points.

Service Switching Points (SSP)

Service Switching Points (SSP) are telephone switches that are interconnected to each other by SS7 links. The SSPs perform call processing on calls that originate, tandem, or terminate at that site. As part of this call processing, the SSP may generate SS7 messages to transfer call-related information to other SSPs, or to send a query to a Service Control Point for instructions on how to route a call.

Signal Transfer Points (STP)

Signal Transfer Points (STP) are switches that relay messages between network switches and databases. Their main function is to route SS7 messages to the correct outgoing signaling link, based on information contained in the SS7 message address fields.

Service Control Points (SCP)

Service Control Points (SCP) contain centralized network databases for providing enhanced services. The SCP accepts queries from an SSP and returns the requested information to the originator of the query. For example, enhanced 800 service uses an SCP database to determine the routing on 800 calls. When an 800 call is initiated by the user, the originating SSP sends a query to an 800 database requesting information to the SSP originating the query and the call proceeds.

SS7 Reliability

To meet the stringent reliability requirements of public telecommunications networks, a number of safeguards are built into the SS7 protocol:

- STPs and SCPs are normally provisioned in mated pairs. On the failure of individual components, this duplication allows signaling traffic to be automatically diverted to an alternate resource, minimizing the impact on service.

- Signaling links are provisioned with some level of redundancy. Signaling traffic is automatically diverted to alternate links in the case of link failures.

- The SS7 protocol has built-in error recovery mechanisms to ensure reliable transfer of signaling messages in the event of a network failure.

ISDN Access Protocol

The Integrated Services Digital Network (ISDN) protocol was designed to provide advanced control messages for the end users. While the SS7 system was designed to provide an internationally standardized, general-purpose signaling system; it was not intended to be used as the signaling standard for access to the telephone network from PBXs or from telephone sets. To satisfy this latter need, the ISDN-AP (Integrated Services Digital Network – Access Protocol) has been developed.

Figure 1.3 shows how the ISDN access protocol can be interconnected with the SS7 system at the telephone switch (SSP) through the user of an ISDN-AP/SS7 Interface. When working together, SS7 and the ISDN-AP provide the end-to-end signaling required to deliver enhanced features to users. As an interim step, some telephone exchange carriers use proprietary access signaling to provide enhanced services.

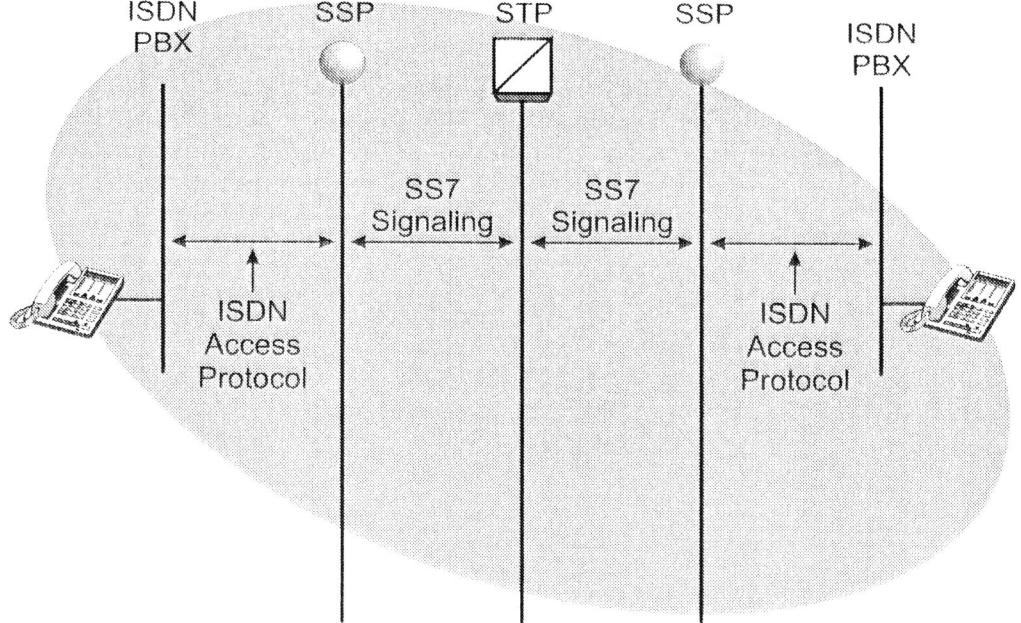

Figure 1.3. ISDN-AP and SS7 Interface

SS7 Mapped Onto the OSI Layer Model

The Open Systems Interconnection (OSI) standard layer model was developed by the International Standards Organization (ISO) and the CCITT. The OSI model helps to standardize the inter-connection of computers and data terminals to their applications, regardless of their type or manufacturer. The protocols specify seven layers: physical, link, network, transport, session, presentation, and application. Each layer performs specific functions for data exchange and is independent of the other layers.

Figure 1.4 shows how the SS7 layers compare to the OSI 7-Layer Reference Model. The bottom half of the SS7 protocol consists of the Message Transfer Part (MTP). There are three levels to the MTP: Level 1 corresponds to the OSI Layer 1 (Physical Layer): Level 2 corresponds to OSI Layer 2 (Data Link Layer): and, Level 3 corresponds to the bottom of OSI Layer 3 (Network Layer). The upper half of the SS7 protocol consists of several parts. The SS7 Signaling Connection Control Part (SCCP) corresponds to the top of OSI Layer 3. The ISDN-User Part (ISDN-UP) maps onto OSI layer 3 as well, and, in addition, it maps onto Layer 4 (Transport Layer), Layer 5 (Session Layer), Layer 6 (Presentation Layer), and Layer 7 (Application Layer). The Transaction Capabilities Application Part (TCAP), the Application Service Elements (ASE), and the Operations, Maintenance and Administration Part (OMAP) of the SS7 protocol all map onto OSI Layer 7 as well.

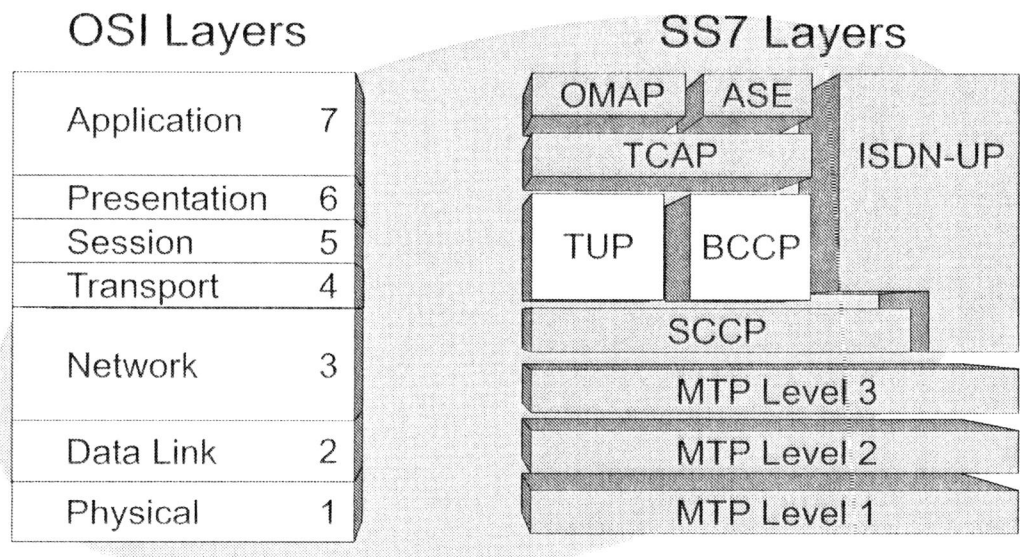

Figure 1.4. Comparison of the SS7 Protocol Layers and the OSI Model Layers

Chapter 2

Message Transfer Part (MTP)
Level 1

The Message Transfer Part (MTP) is the functional part of a common channel signaling system that transfers signaling messages as required by all the users. The message transfer part also contains, for example, error control and signaling security. The Message Transfer Part (MTP) Level is known as the signaling data link. It is equivalent to the OSI model Physical Layer (Layer 1).

Figure 2.1 displays where the MTP Level 1 part is within the SS7 system. This diagram shows that MTP1 is responsible for the physical transport of information on the signaling links that connect SS7 components. MTP1 defines the physical, electrical, and functional characteristics of the signaling links.

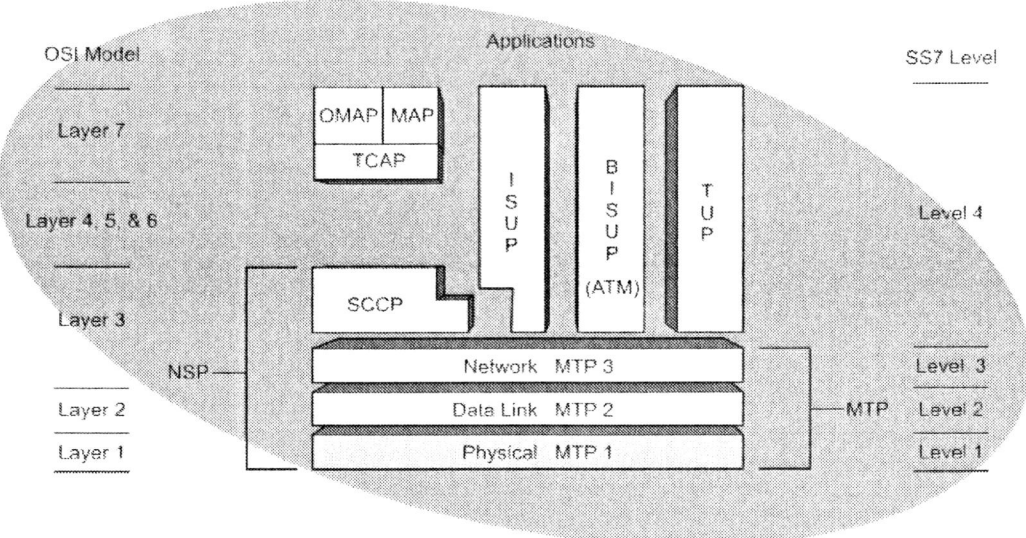

Figure 2.1, MTP Level 1 in the SS7 Layers

A signaling data link is a bi-directional transmission path that allows for the transfer of signaling messages. The transfer of signaling messages over a communication link must transfer at the same data rate in either direction. SS7 signaling data links are capable of operation over terrestrial and satellite transmission links. The signaling terminal at each end of the signaling data link contains the MTP Level 2 functionality for transmitting and receiving SS7 messages.

Digital Signaling Link

A digital signaling link is composed of a digital transmission channel connecting two digital switches or network nodes that provide an interface to the signaling terminals.

Figure 2.2 shows how an SS7 system uses a digital signaling link to transfer messages. To transfer data messages on a digital line, data communications equipment (DCE) are attached at the end of each digital switch to convert the messages from digital form unique to the switching system to the SS7 form. The DCE buffers and converts messages between the signaling link and the switching system's signaling terminal.

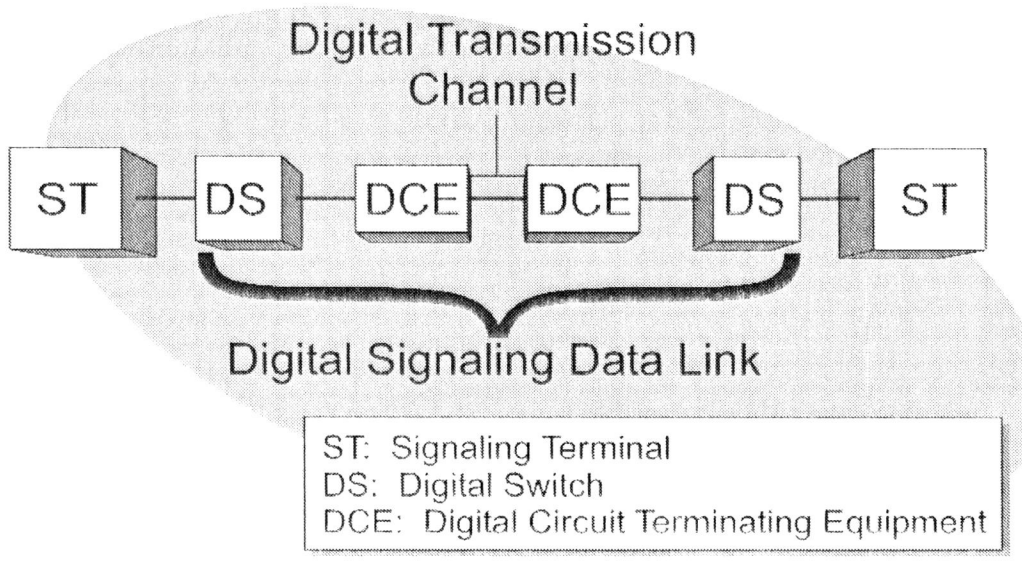

Figure 2.2. SS7 Digital Signaling Data Link

Digital signaling links are often used on newer switching systems. Because signaling messages are relatively small in content size, the common data transmission rates for digital signaling links is 56 kilobits/sec (Kb/s) or 64 Kb/s. The minimum signaling rate for call control applications on the SS7 network is 4.8 Kb/s.

Analog Signaling Link

An analog signaling data link is made up of a voice-frequency analog transmission channel connecting digital switches that provide an interface to the signaling terminals.

Figure 2.3 shows how an SS7 system can use an analog signaling links to transfer SS7 data messages. To transfer data messages on an analog line, modems are attached at the end of each digital switch to convert the messages from digital form to analog form. This permits the transfer of digital information on the analog communication line. Analog signaling links are often used on older switching systems and they usually have signaling transmission rates that are lower than digital signaling links.

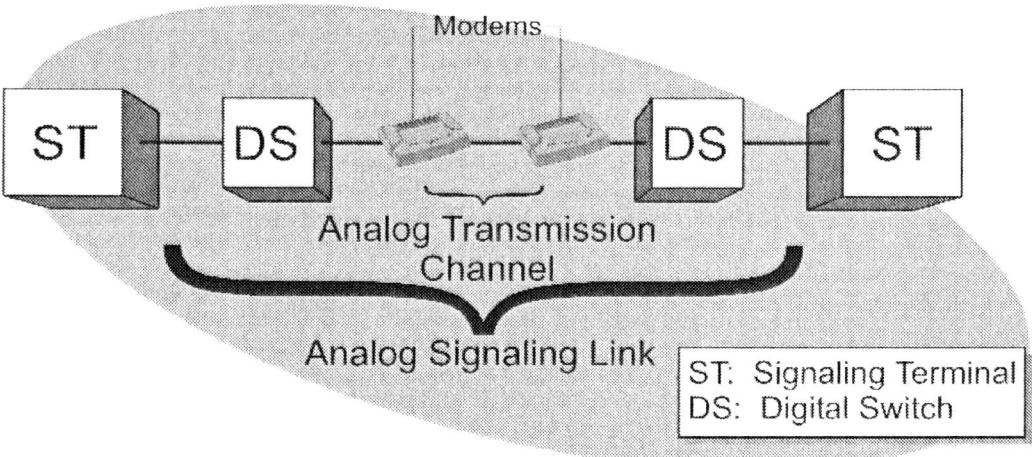

Figure 2.3, SS7 Analog Signaling Data Link

Chapter 3

Message Transfer Part (MTP) Level 2

The Message Transfer Part (MTP) Level 2, together with MTP Level 1, provides a signaling link for reliable transfer of signaling messages between two directly connected signaling points. The data link includes all equipment and signals in the connection.

Figure 3.1 displays where the MTP Level 2 functional part is located within the SS7 system. This diagram shows that MTP2 is responsible for the setup (initial alignment), maintaining (flow control), and disconnection (error monitoring) of links between SS7 signaling points.

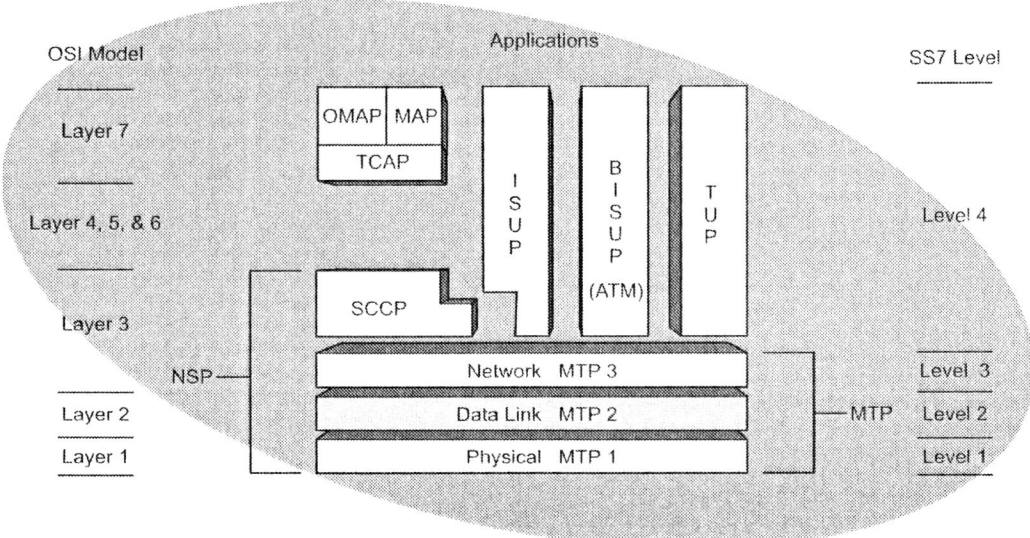

Figure 3.1, MTP Level 2 of the SS7 Layers

Message Formats

There are three types of signal unit (SU) message formats. They are differentiated by the value contained in the length indicator (LI) field. The signal unit field types include:

F (flag):   The beginning and end of a signal unit are indicated by a unique 8-bit pattern, called the flag, which does not appear elsewhere in the signal unit. Measures are taken to ensure that the flag pattern is not imitated elsewhere in the SU. The flag pattern is 01111110.

CK (cyclic redundancy check):   The CK is a 16-bit checksum transmitted with each signal unit. If the checksum does not match at the receiving signaling point, the SU is considered to have errors and is discarded.

SIF (signaling information field):   This field contains the routing and signaling information of the message.

SIO (service information octet):   This octet is made up of the service indicator and the sub-service field. The service indicator is used to associate the signaling message with a particular MTP user at a signaling point, for instance, the layers above the MTP level. The sub-service field contains the network indicator that is used to differentiate between national and international calls, or between different routing schemes within a single network.

BSN (backward sequence number):   The BSN field is used to acknowledge message signal units which have been received from the remote end of the signaling link. The BSN is the sequence number of the signal unit being acknowledged.

BIB (backward indicator bit):   The BIB is used in error recovery.

FSN (forward sequence number):   The FSN is the sequence number of the signal unit in which it is being carried.

FIB (forward indicator bit): The FIB is used in error recovery.

LI (length indicator):   The LI field indicates the number of octets that follow the LI field and precede the DK field.

- LI = 0   Fill-in signal unit (FISU)
- LI = 1 or 2   Link status signal unit (LSSI)
- 2 < LI < 63   Message signal unit (MSU)

Message Signal Units (MSU)

Message Signal Units carry signaling information for call control, network management, and maintenance I the signaling information field. For example, messages from the Signaling Connection Control Part (SCCP), the ISDN-User Part (ISDN-UP), and the Operations, Administration and Maintenance Part (OMAP) are transferred over the signaling link in the signaling information field of variable length MSUs.

Figure 3.2 shows the packet structure of the MSU. This diagram shows that the MSU has a variable length SIF field that allows the MSU to carry many types of signaling packets. These include SCCP, ISDN-User Part, and OMAP messages.

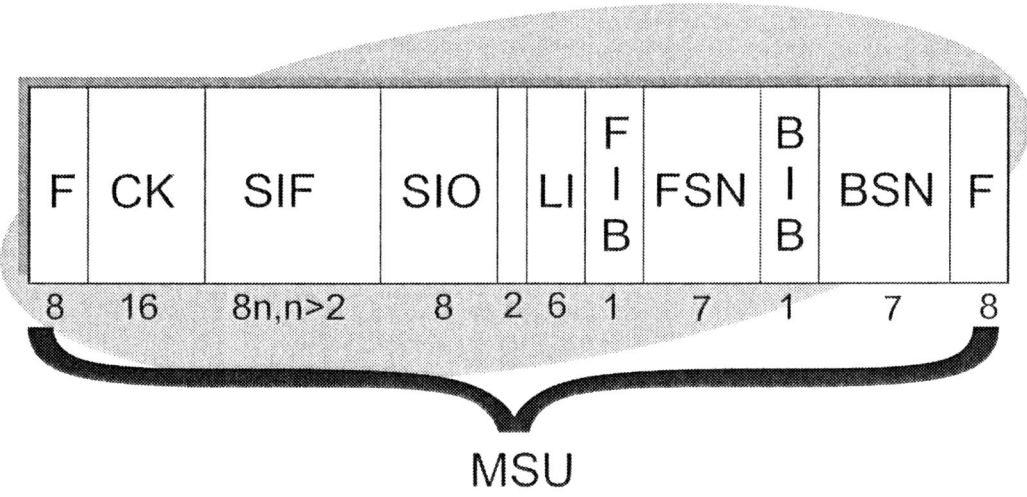

Figure 3.2, Message Signal Unit (MSU)

Link Status Signal Unit (LSSU)

The Link Status Signal Unit (LSSU) messages provide link status indications to the remote end of the signaling link. Some examples of status indications are: normal, out of alignment, out of service, emergency status.

Figure 3.3 shows the packet structure of the LSSU. This diagram shows that the SIF value is limited to 8 of 16 bits. This simple structure is used to uniquely identify link test messages that are sent during alignment, out of service, or emergency status conditions.

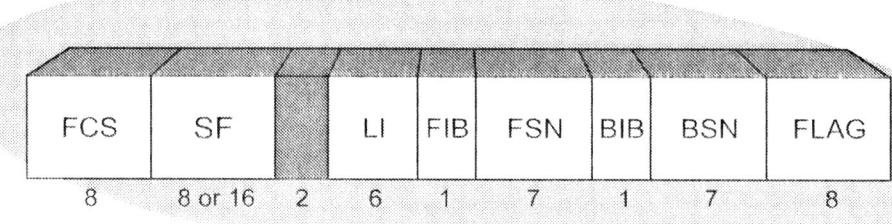

Figure 3.3. Link Status Signal Unit (LSSU)

Fill-in Signal Unit (FISU)

The Fill-in Signal Unit (FISU) messages are transmitted when no MSUs or LSSUs are being transmitted, allowing the SS7 network to receive immediate notification of signaling link failure.

Figure 3.4 shows the packet structure of the FISU. This diagram shows the FISU packet has a single format structure. This sending of this simple packet structure allows the SS7 network to quickly detect if the signaling link experiences a failure when no other messages are being sent.

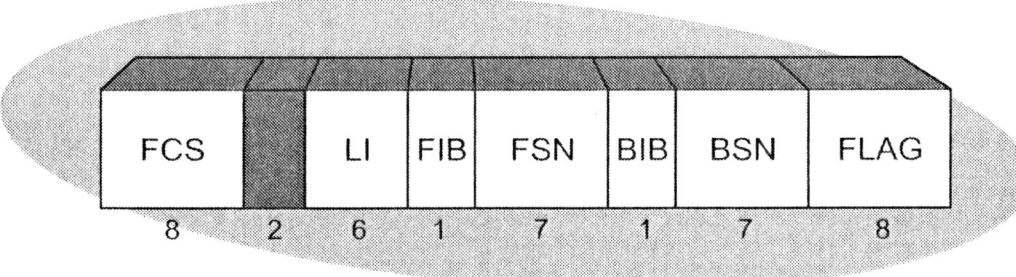

Figure 3.4, Fill-in Signal Unit (FISU)

Link Control Procedures

The link control procedures include the setup (initial alignment), error monitoring, error correction, and flow control.

Initial Alignment Procedures

Initial alignment procedures (IAP) occur when a signaling link is activated for the first time or restored after a link failure. Two alignment procedures are provided, normal and emergency. A "proving state" is included in both procedures to measure error rates, ensuring that a reliable link is established.

Signaling Link Error Monitoring

Two different types of signaling link error rate monitors are provided to estimate signaling link error rates: the signaling unit error rate monitor (SUERM), and the alignment error rate monitor (AERM).

The SUERM is employed while signaling links are in service. The SUERM provides a fault indication to MTP Level 3 when error thresholds are exceeded. The AERM is used during the initial alignment procedure providing state.

Error rate thresholds are different for each of the two error rate monitors. The error thresholds are based on a combination of the percent message error rate and the length of time over which errors are occurring. For example, at 100% error rate, the threshold would be reached in 128 ms, whereas with a lower error rate, the threshold would not be reached until a longer time had passed.

Error Correction

Two forms of error correction are provided, the Basic Method and the Preventive Cyclic Retransmission Method. Both methods are designed to eliminate the possibility of missed, duplicated, or out-of-sequence messages.

The Preventive Cyclic Retransmission method is used on signaling links such as satellite links that have large propagation delays.

The basic method of error correction is accomplished by retransmitting those MSUs that were not correctly received in order by the destination signaling point.

Normally, the destination signaling point will reply to a transmitted MSU with a positive acknowledgement. Reception of a positive acknowledgement at the origination signaling point confirms the successful transmission of that MSU. If the exchange sending back the acknowledgement has messages to be sent back at the same time, the acknowledgement can be contained within the MSU carrying the message. However, if a negative acknowledgment is returned from the destination signaling point, the originating signaling point will retransmit the MSU and all subsequent MSUs.

Figure 3.5 shows how basic error correction is performed on an SS7 link:

Exchange A transmits an MSU with Forward Sequence Number (FSN) = 4.

Exchange B acknowledges the successful receipt of the MSU from Step 1 by setting the Backward Sequence Number (BSN) = 4 in the FISU it sends to exchange A.

Exchange A has two MSUs to transmit. FSN = 5 and FSN = 6 are selected and transmitted in order. In this example, the MSU with FSN = 5 was corrupted by transmission errors. Exchange B receives the MSU with FSN = 6 correctly.

Exchange B sends a negative acknowledgement to exchange A indicating that the MSU with FSN = 4 was the last MSU successfully received in order. The negative acknowledgement is indicated by toggling the value of the Backward Indication Bit (BIB).

Exchange A now retransmits the MSUs with FSN = 5 and FSN = 6, which are received successfully by exchange B.

Exchange B now acknowledges these MSUs by replying with an FISU with BSN = 6. The FISU that serves as the acknowledgement for MSU with FSN = 6 also serves as an acknowledgement for all previous unacknowledged MSUs as well (in this case, MSU with FSN = 5). An exchange can send up to 127 MSUs before requiring an acknowledgement from the remote end.

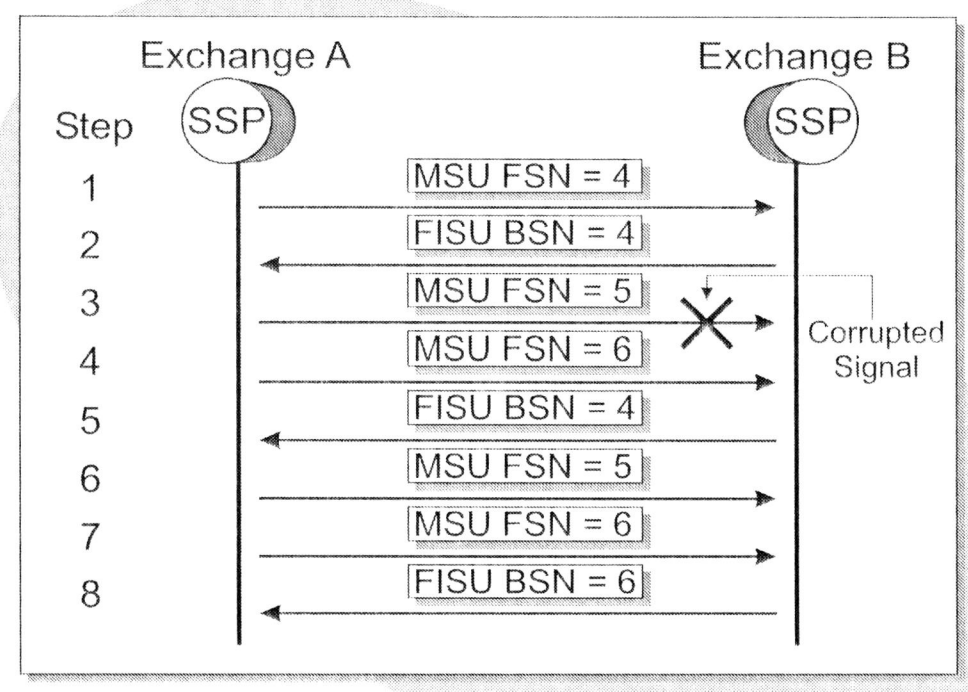

Figure 3.5. SS7 Link Basic Error Correction

The local signaling point cyclically retransmitting all the MSUs sent but not yet acknowledged by the remote signaling point accomplishes the preventive method of error correction. When there are no new MSUs or LSSUs to be transmitted, all the MSUs that have not been positively acknowledged are retransmitted cyclically.

Figure 3.6 shows how preventive error correction (Cyclic Retransimission Correction) method can be used on an SS7 link:

Exchange A transmits an MSU with FSN = 4.

Exchange B acknowledges the successful receipt of the MSU from Step 1 by returning and FISU with BSN = 4 to exchange A.

Exchange A sends two additional MSUs to exchange B

Exchange A has no additional MSUs to transmit and has not received acknowledgement from the MSUs sent in Step3.f Exchange A then retransmits the MSUs with FSN = 5 and FSN = 6.

Exchange B acknowledges the MSU with FSN = 6, confirming the receipt of MSU with FSN = 5 as well.

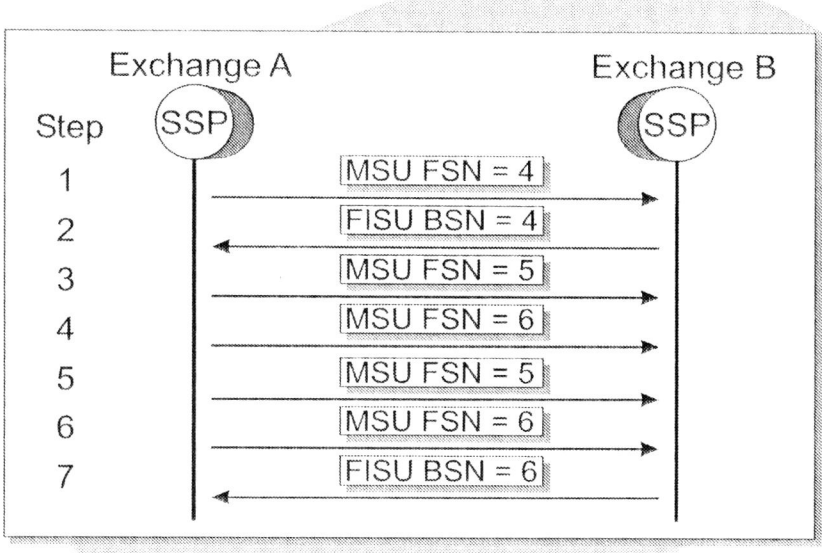

Figure 3.6. SS7 Link Preventive Error Correction

Flow Control

Flow control procedures are initiated on a signaling link when congestion is detected at either end of the signaling link. Congested conditions may be due to processor outage or link failure anywhere on the network.

The end of the signaling link initiating flow control withholds both positive and negative acknowledgements, and sends a busy status indication to the remote end of the signaling link. If congestion persists, the remote end removes the signaling link from service and indicates an emergency rerouting procedure.

Chapter 4

Message Transfer Part (MTP) Level 3

The Message Transfer Part (MTP) Level 3 provides the functions and procedures related to message routing and network management. MTP Level 3 handles these functions, assuming that signaling points are connected with signaling links as described in MTP Level 1 and Level 2.

Figure 4.1 displays where the MTP Level 3 functional part is located within the SS7 system. This diagram shows that MTP3 is responsible for the setup of a connection through the SS7 network, maintaining the connections, and managing the reconfiguration of connections. This diagram shows that MTP3 also is responsible for communicating and converting messages into the formats used by upper layer applications.

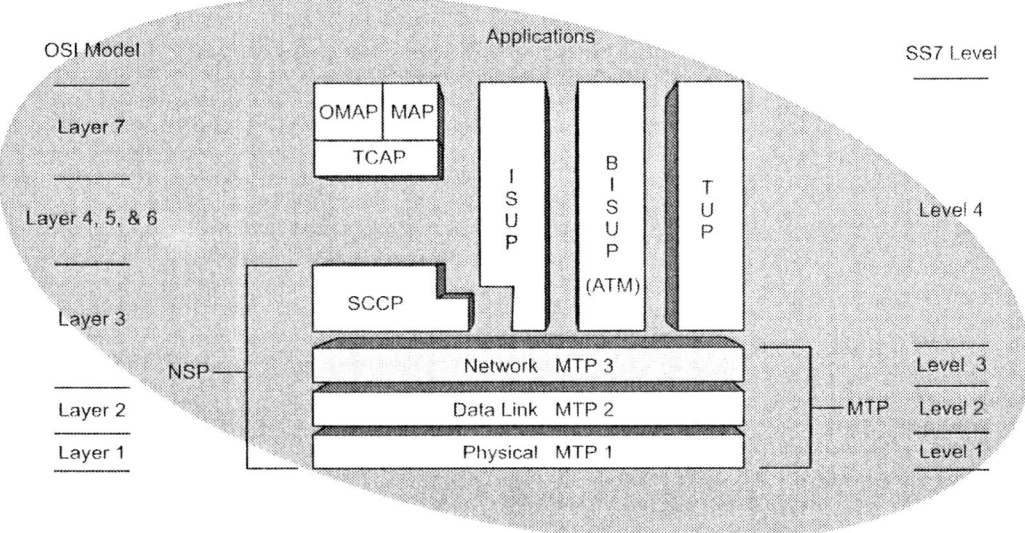

Figure 4.1. MTP Level 3 of the SS7 Layers

The MTP Level 3 functions can be divided into two basic categories: signaling message handling functions, and signaling network management functions.

Figure 4.2 shows how SS7 MTP3 level is responsible for processing messages and that some of the messages are used for network management functions. At each signaling point in the signaling network, message routing, discrimination and distribution functions are performed.

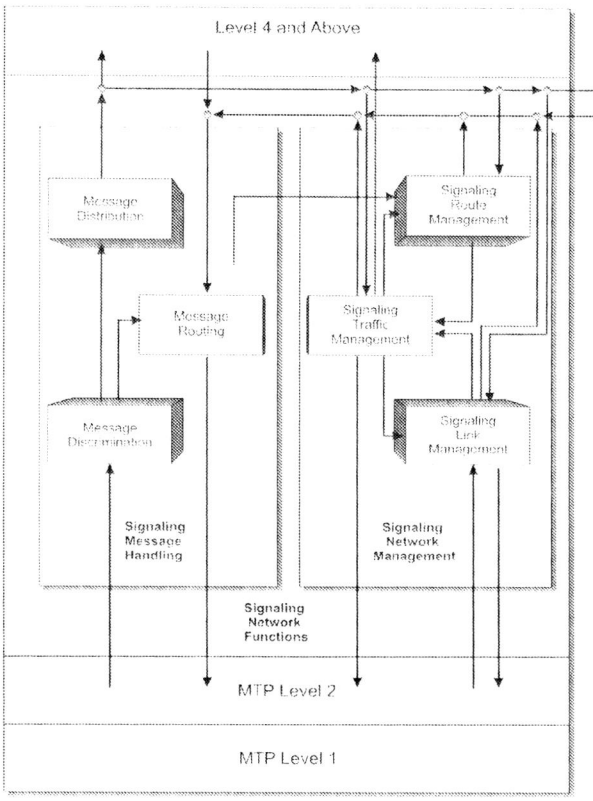

Figure 4.2. Signaling Network Functions

The signaling network management functions provide the actions and procedures required to activate and maintain signaling service, and to restore normal signaling conditions in the event of disruption in the signaling network, either in signaling links or at signaling points.

Signaling Message Handling

Signaling message handling is based on the routing label contained in the signaling information field (SIF) of the message signal units. Link status and fill-in signal unit messages only travel between two signaling points, therefore neither of these messages contains routing labels.

Figure 4.3 shows a routing label that is contained within the SIF field of the MSU. The routing label identifies the originating and destination address for the message. In some cases, the service information octet (SIO) is used for routing purposes as well.

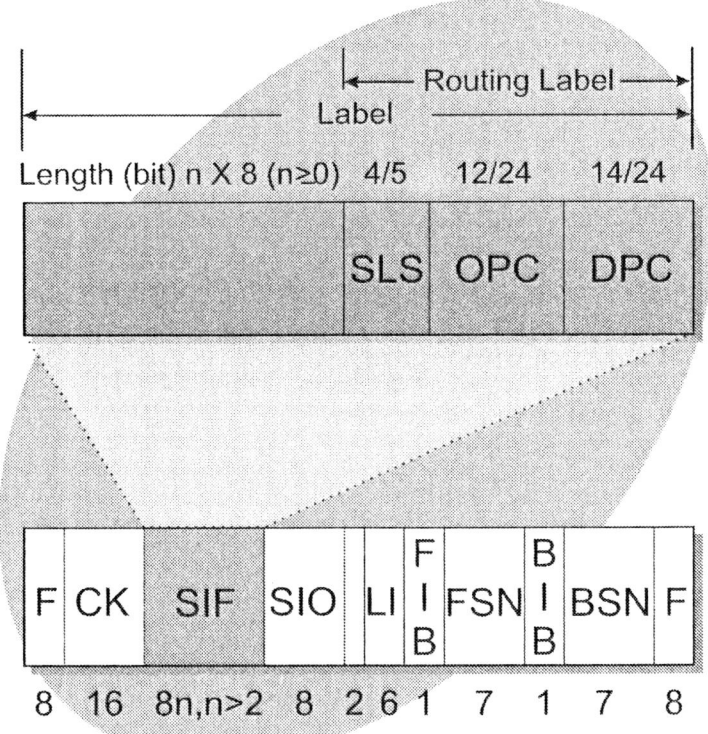

Figure 4.3, Routing Label Contained Within SIF of the MSU

Signaling Information Field: Routing Label

The standard routing label assumes that each signaling point in a signaling network is allocated a code according to a labeling code plan that is unambiguous in its domain. The routing label includes:

> The originating and destination point codes; and,
> The signaling link selection code.

The originating point code (OPC) indicates the originating point of the message, while the destination point code (DPC) identifies the destination of the message.

The signaling link selection (SLS) field is used for load sharing when two or more links connect adjacent signaling points. Each signaling link is assigned as SLS value. Messages are routed over that signaling link when the MTP 3 sets the SLS field value equal to that of the signaling link.

Figure 4.4 shows how the SLS process is used to obtain the signaling link code (SLC) that identifies the signaling channel that is used between SS7 network nodes. The SLC code is a pre-assigned code that identifies a logical channel on a physical link. All links are assigned a SLC code and a terminal number.

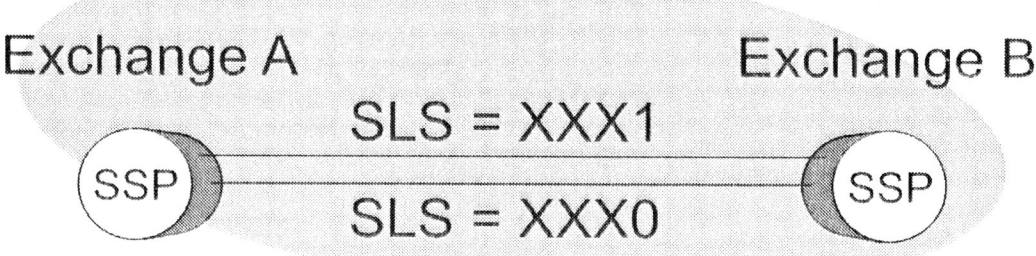

Figure 4.4. Signaling Link Selection (SLS)

Message Routing

To route messages, each signaling point has routing information which allows it to determine the signaling link over which a message has to be sent based on information contained in DPC and SLS fields.

Typically, more than one signaling link may be used to carry messages to a given DPC; the selection of the particular signaling link is made by means of the SLS field, thus effecting load sharing.

Message Discrimination

The DPC field of the received message is examined by the discrimination function. The message discriminator determines if the message has a DPC of its device(s) or if message is to be sent to another location and routes the message accordingly.

Figure 4.5 shows how a message is processed at the receiving signaling point in an SS7 system. If the message is designated for the receiving signaling point, the message is delivered to the message distribution functions.

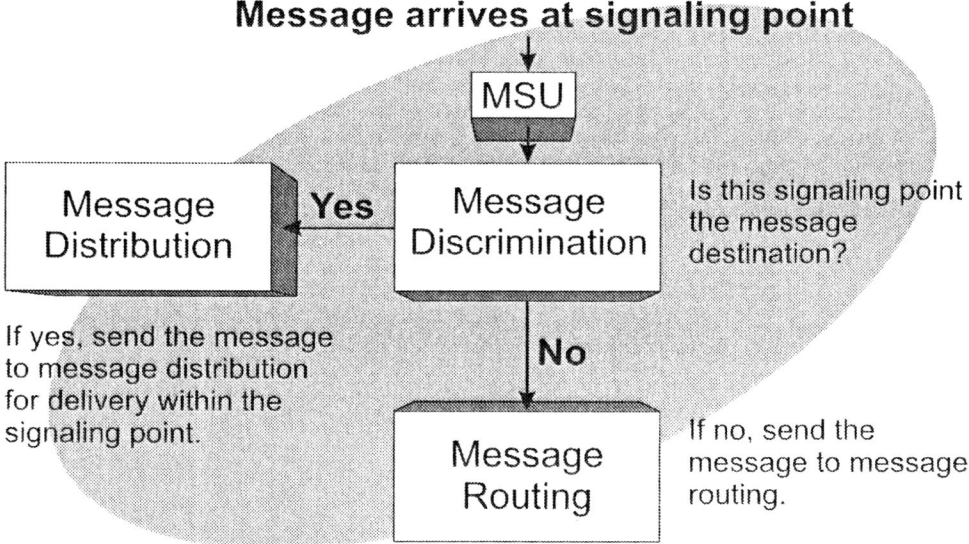

Figure 4.5. Destination Point Code

If the DPC is not that of the receiving signaling point, and if the receiving signaling point has the transfer capability, the message is directed to the message routing function.

Message Distribution

The message distribution function of the MTP3 layer transfers the message to the appropriate user part. It determines which user part for distribution by reviewing the SIO field contained within the MSUs.

Figure 4.6 shows an example of some how the MTP3 layer is used to distribute messages to the upper layer user parts. This diagram shows that messages may be transferred to different types of service indicators including SCCP, ISDN User Part, Signaling Network Management, and Signaling Network Testing.

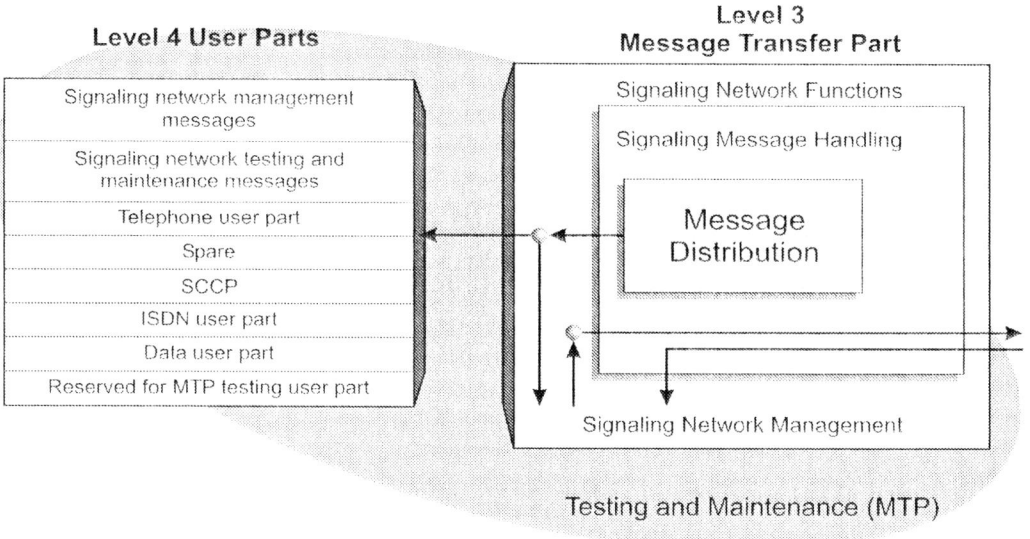

Figure 4.6. MTP3 Message Distribution

Signaling Network Management

The purpose of the signaling network management functions is to activate new signaling links, to maintain signaling service, to control traffic in case of congestion, and to provide reconfiguration of the signaling network in the case of failures.

In the case of failures, traffic will be rerouted around the failed component if possible, and new signaling links may be activated. Congestion generally results in the change in status of the affected signaling links and routes from an "available" state to an "unavailable" state.

The signaling network management functions are divided into three categories:

> Signaling link management;
> Signaling route management; and,
> Signaling traffic management.

Signaling Link Management

The signaling management function manages locally attached signaling links. The function is responsible for maintaining predetermined link set capabilities by establishing link sets and initiating action to activate additional links in the event of signaling link failures.

Basic signaling link management capabilities are defined for link activation, deactivation, restoration, and link set activation. Additional capability sets are provided for automatic activation of backup signaling links.

The signaling link management function is carried out by the following procedures: link set activation, signaling link deactivation, signaling link activation, and signaling link restoration.

Link Set Activation

Normal or emergency link set activation procedures are used to establish link sets having no active signaling links.

Normal activation is initiated when a link set is being activated for the first time or when a link set is being restarted and the situation is deemed not to be an emergency. The MTP Level 2 normal Initial Alignment Procedures (IAP) is used on each signaling link in parallel.

Emergency activation is initiated when an immediate re-establishment of the link set is required. The MTP Level 2 emergency IAPs are used.

Signaling Link Deactivation

Active signaling links may be taken out of service for two reasons:

1. The quality of active signaling links exceeds the predetermined quantity for that link set.
2. Maintenance activities may require a link to be manually deactivated for testing.

Signaling Link Activation

This procedure is used to activate signaling links which are being activated for the first time, or which were previously taken out of service.

Signaling Link Restoration

On detection of a signaling link failure, the signaling link restoration procedure is initiated. This in turn initiates the MTP Level 2 IAP. On successful completion of the IAP, the link is brought into service.

Ti the IAP fails, a new IAP is repeated until the link is restored, or manual intervention takes place.

Signaling Route Management

The signaling route management functions are used to exchange signaling route availability information between signaling points. Information exchange is accomplished using transfer prohibited, transfer restricted, transfer allowed signaling route set test, transfer controlled, and signaling route set congestion test procedures.

Transfer Prohibited

The transfer prohibited procedure is performed at a signaling point acting as a Signal Transfer Point (STP) for messages relating to a given destination when it has to notify one or more adjacent signaling points that they must no longer route messages for the given destination via that STP.

Figure 4.7 shows an example of an SS7 link transfer prohibited operation:

> On the failure of links D-F and D-E, D no longer has access to F. D indicates this condition to B and C by sending them transfer prohibited messages identifying that F can no longer be reached via D.
>
> B and C will then indicated the forced rerouting procedure to reroute signaling traffic destined for F via E.

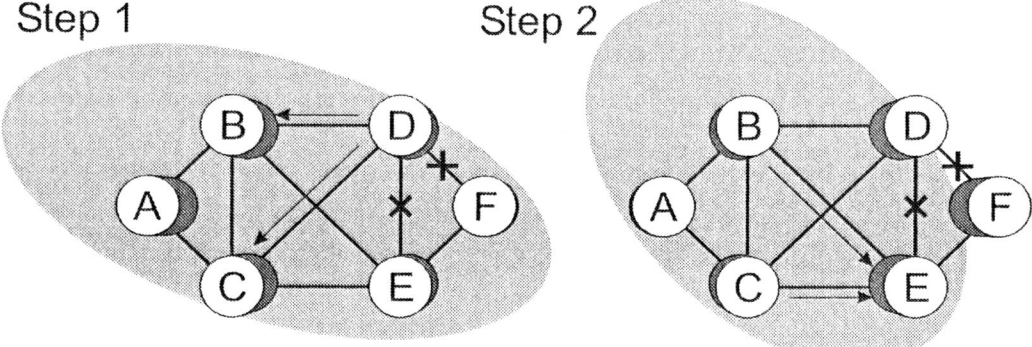

Figure 4.7. SS7 Link Transfer Prohibited

Transfer Restricted

The transfer restricted procedure is performed at an STP when it has to notify one or more adjacent signaling points that they should not, if possible, route traffic toward a given destination through that signaling point. Transfer restricted messages are sent under certain link failure and congestion situations.

Figure 4.8 shows an example of an SS7 link transfer restricted operation:

> On a failure of link A-B, a changeover procedure is initiated between A and B. Signaling traffic between A and F may take the following path: A-C-B-D-F (highlighted in the illustration). Signaling traffic management may decide that alternate routing over path A-C-D-F is more efficient.

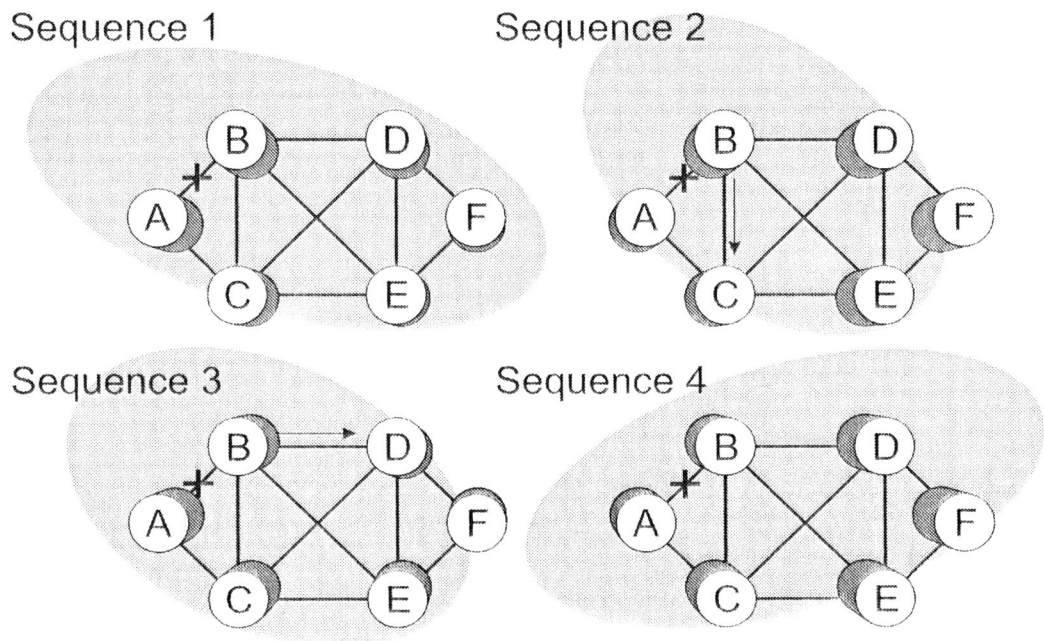

Figure 4.8. SS7 Link Transfer Restricted

If so, B will send a transfer restricted message to C indicating that C should, if possible, reroute traffic destined for F. In this case, C would initiate the controlled rerouting procedure and route traffic destined for F via D.

Similarly, B will send a second transfer restricted message to D indicating that D should, if possible reroute traffic destined for A. In this case, D would initiate the controlled rerouting procedure and route traffic destined for A via C.

The new route for messages going between A and F is highlighted in the illustration.

Transfer Allowed

The transfer allowed procedure is performed at an STP when it has to notify one or more adjacent signaling points that they may route traffic toward a given destination through that signaling point.

Figure 4.9 shows an example of an SS7 link transfer allowed operation:

A previous failure of links D-D and D-F would have resulted in B and C rerouting traffic for F via E.

On recovery of link D-F, D will send a transfer allowed message to B and C identifying that F could now be reached via D.

If D is the primary route for sending traffic from B and C to F, then B and C may than initiate the controlled rerouting procedure to reroute the signaling traffic destined for F via D.

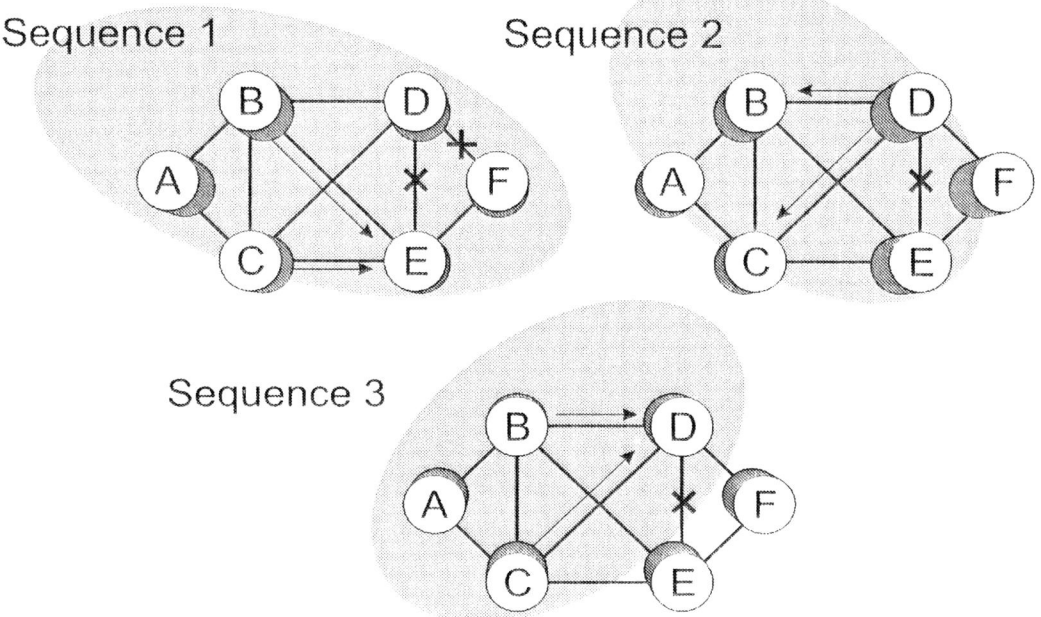

Figure 4.9. SS7 Link Transfer Allowed < ag_SS7_Link_Transfer_Allowed>

Signaling Route Set Test

This test is performed at signaling points to query whether signaling traffic towards a certain destination may be routed via an adjacent STP. This procedure is activated following receipt of a transfer prohibited or transfer restricted message from an adjacent signaling point.

Figure 4.10 shows an example of the operation for signaling route set testing in the SS7 system:

> A previous failure of links D-E and D-F would have resulted in B and C rerouting traffic for F via E.

In this case, B and C send Signaling Route Set Test messages to D, requesting the status of the route to F. This occurs every 30 to 60 seconds until a transfer allowed message is received from D. (A transfer allowed message indicates that the destination has become available.)

Point D responds to the Signaling Route Set Test message with a transfer allowed, restricted or prohibited message as dictated by the current status.

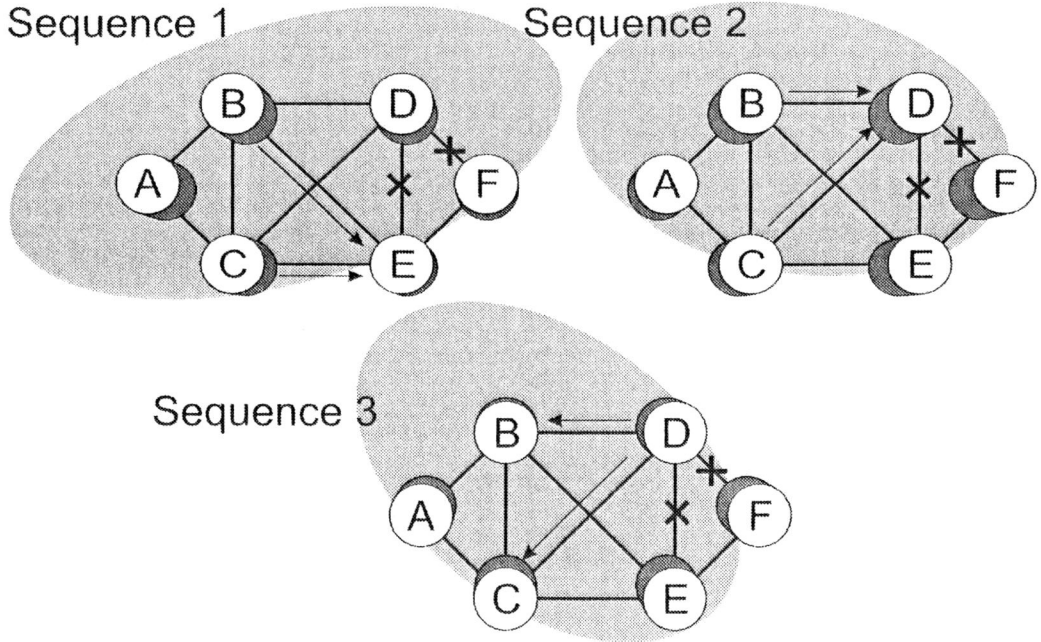

Figure 4.10. Signaling Route Set Test

Transfer Controlled

The transfer controlled procedure is used to notify adjacent signaling points of congestion at a signaling point or the congestion status of a signaling route. The action taken on receipt of a transfer controlled message depends on the implementation of the network operator.

Signaling Route Set Congestion Test

This test is performed at a signaling point to determine the congestion status of a specific destination. A signaling point may query the signaling route status at other signaling points by the signaling route set-test procedure.

Congestion information is transmitted and requested by the transfer controlled message and signaling route set congestion test messages.

Figure 4.11 shows an example of how signaling route set congestion testing can be performed on an SS7 system.

> B sends the congestion test message to D querying the congestion status of the route B-D-F.

> D may respond with a transfer controlled message indicating the congestion status level. If D does not respond within a set time interval, then B will resend the congestion test message to D. Specific details of this procedure are network-dependent.

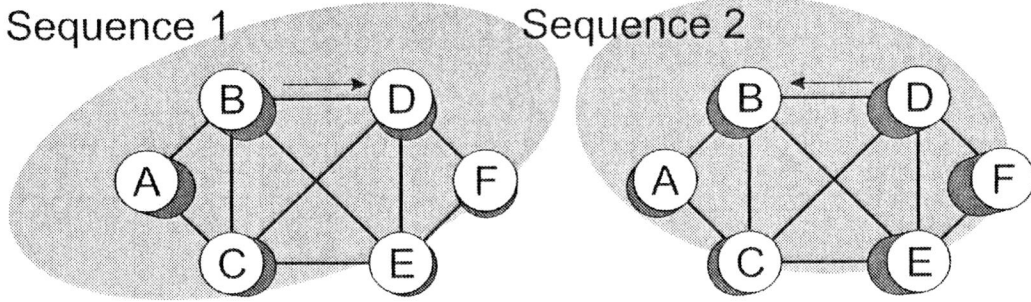

Figure 4.11. Signaling Route Set Congestion

Signaling Traffic Management

The signaling traffic management functions are used to divert signaling traffic from a link or route to one or more different links or routes, or to temporarily slow down signaling traffic in the case of congestion at a signaling point.

The diversion of traffic in cases of unavailability, or availability, or restriction of signaling links and routes is typically made by means of the following basic procedures.

Signaling Point Restart

This procedure is used when a signaling point first becomes available, that is, when it establishes an active signaling link with another signature point in the SS7 network.

When the first signaling link of a signaling link set becomes active, signaling traffic destined for the remote end of the link set is restarted. As signaling links become active, transfer allowed, restricted, and prohibited messages are exchanged between the restarting and adjacent signaling points.

Management Inhibiting

A management inhibiting procedure makes a signaling link unavailable to user part traffic, but still allows maintenance and testing traffic to be carried out on the signaling link. This may be necessary where reliability problems are being experienced on the signaling link.

A request to inhibit a signaling link is not carried out if it means that an accessible destination would become inaccessible or if congestion was currently a problem in that part of the SS7 network.

Signaling Traffic Flow Control

The signaling traffic flow control procedure restricts signaling traffic at its source when network failures or congestion occur. A number of conditions will trigger flow control action:

> Signaling Route Set Unavailability: When the MTP determines that there is no signaling route available for a particular destination, the MTP informs the local user parts that signaling traffic destined for that destination cannot be transferred via the SS7 network. The local user parts take appropriate steps to stop placing signaling traffic for that destination on the network.

> Signaling Route Set Availability: When a transfer allowed message informs an MTP that a signaling route is not available to a previously inaccessible signaling point, the MTP informs the local user parts that signaling traffic destined for that particular signaling point can be transferred via the SS7 network. The local user parts take appropriate steps to begin transmitting signaling traffic for that destination.

Signaling Route Set Congestion: When the MTP is informed that the signaling route to a particular destination is congested, the MTP informs the local User Parts of the congestion. In networks with congestion priorities, the MTP informs the User Parts of the signaling point's congestion status.

User Part Failure: If an MTP is unable to deliver a received message to a local User Part, the MTP sends a User Part unavailable message to the signaling point originating the message. The originating MTP, on receiving the User Part unavailable message, informs the affected local User Part of the failure at the remote User Part.

Forced Rerouting

The forced rerouting procedure to be used when a signaling route towards a given destination becomes unavailable. The objective of the forced rerouting procedure is to restore, as quickly as possible, the signaling capability toward a particular destination, in a way that minimizes the consequences of failure. Forced rerouting is initiated at a signaling point when a transfer prohibited message, indicating signaling route unavailability, is received.

Figure 4.12 shows example of the operations that can be performed to force rerouting.

In the event that links D-F and D-E fail, D sends transfer prohibited messages to B and C, indicating that they must not route traffic destined for F via D.

Transmission of signaling traffic towards F on the unavailable routes is immediately stopped, and such traffic is stored in forced rerouting buffers. An alternate route is determined according to the rules in Signaling Traffic Management. In this case, both B and C would select the links to E as the alternate route.

As soon as Step 2 is completed, the signaling traffic destined for F is restarted on the alternate route link sets to E, starting with the contents of the forced rerouting buffer.

B and C would send transfer allowed messages to D, indicating to D that messages destined for F could be routed via B and C.

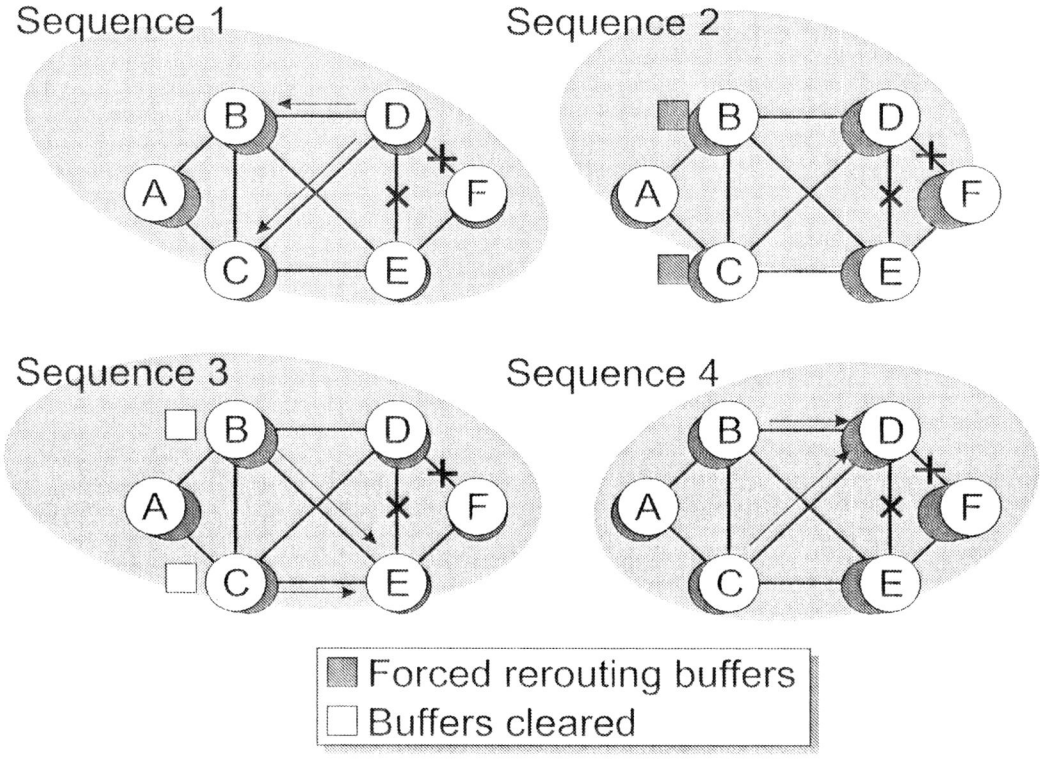

Figure 4.12. SS7 Link Forced Rerouting

Controlled Rerouting

The controlled rerouting procedure is used to restore the optimal signaling route following certain types of changes in the status of signaling links and signaling routes.

Figure 4.13 shows an example of how controlled rerouting can be activated on the receipt of a transfer restricted or a transfer allowed message.

> In the previous example, we saw that with the failure of links D-F and D-E, B and C initiated the forced rerouting procedure to route traffic destined for F via E.

> On the recovery of link D-F, D sends a transfer allowed message to B and C indicating that F can now be reached via D. If D is the primary route for signaling traffic originating at B or C and destined for F, then the controlled rerouting actions are initiated at B and C.

> Transmission of signaling traffic at B and C destined for F via E is stopped and placed into a controlled rerouting buffer. Transfer prohibited messages are sent from B and C to D, indicating to D that D is not to route signaling traffic destined for F via B or C. These transfer prohibited messages are used to ensure that rerouting loops do not occur.

> After a pre-set time interval (usually one second), the signaling traffic contained in the buffers at B and C is transmitted on signaling links to D.

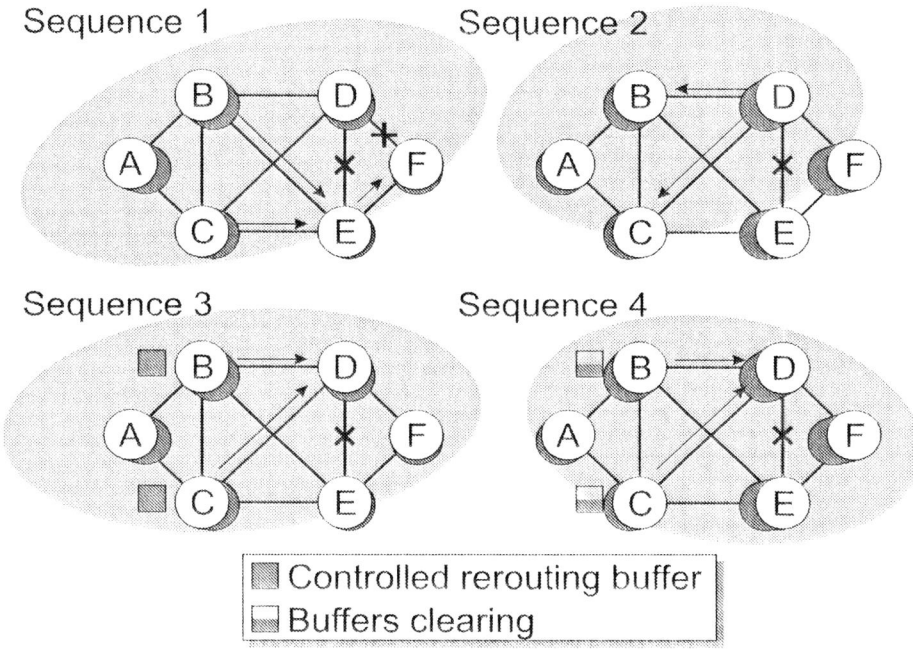

Figure 4.13. SS7 Link Controlled Rerouting

Changeover

The changeover procedure is used to ensure that signaling traffic carried by an unavailable signaling link is diverted to the alternate signaling link(s) as quickly as possible while avoiding message loss, duplication or mis-sequencing. The changeover procedure includes buffer updating, which is performed before reopening the alternative signaling link(s) to the diverted traffic.

Buffer updating consists of identifying all those messages in the retransmission buffer of the unavailable signaling link which were transmitted but have not been received by the far end.

Figure 4.14 shows an example of the operations that can be performed for the changeover process. This example shows that changeover messages are first sent to the existing link, the equipment is this link acknowledges the changeover messages, and a new link is established.

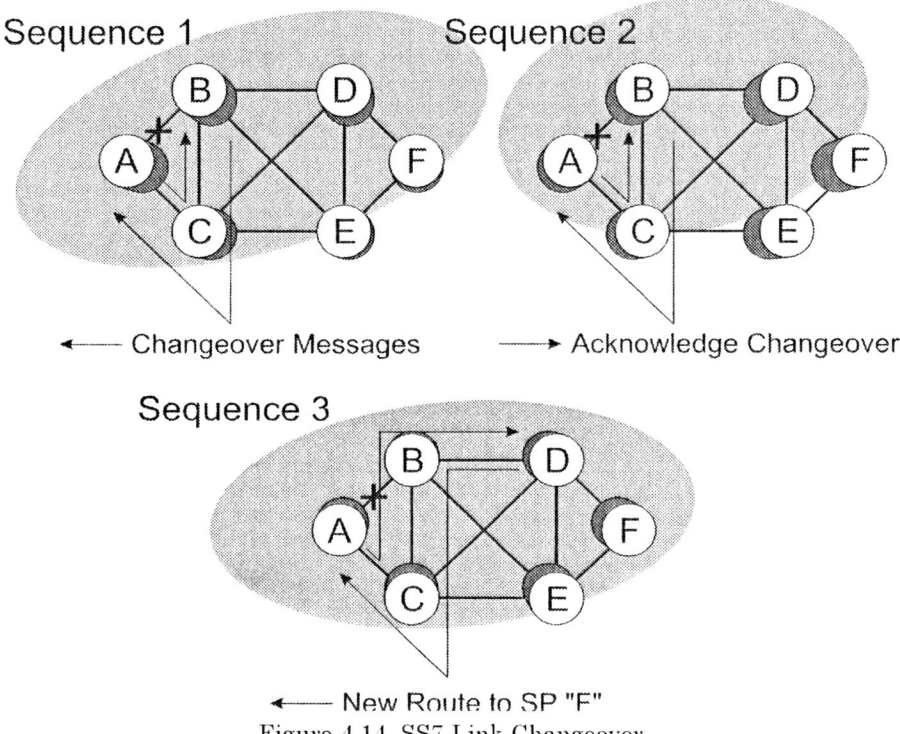

Figure 4.14. SS7 Link Changeover

Changeback

The changeback procedure is used to ensure that signaling traffic is diverted from alternate signaling link(s) back to a link made available as quickly as possible while avoiding message loss.

Figure 4.15 shows an example of the operations that can be performed out for the changeback procedure.

Sequence

In the previous example, A and B initiated the changeover procedure when link A-B failed.

When the A-B signaling link becomes active again, a changeback message is issued by the signaling points at both ends of the link,

Both signaling points respond to changeback declarations with a changeback acknowledgement (ack).

Signaling points A and B immediately start sending traffic on the link made available.

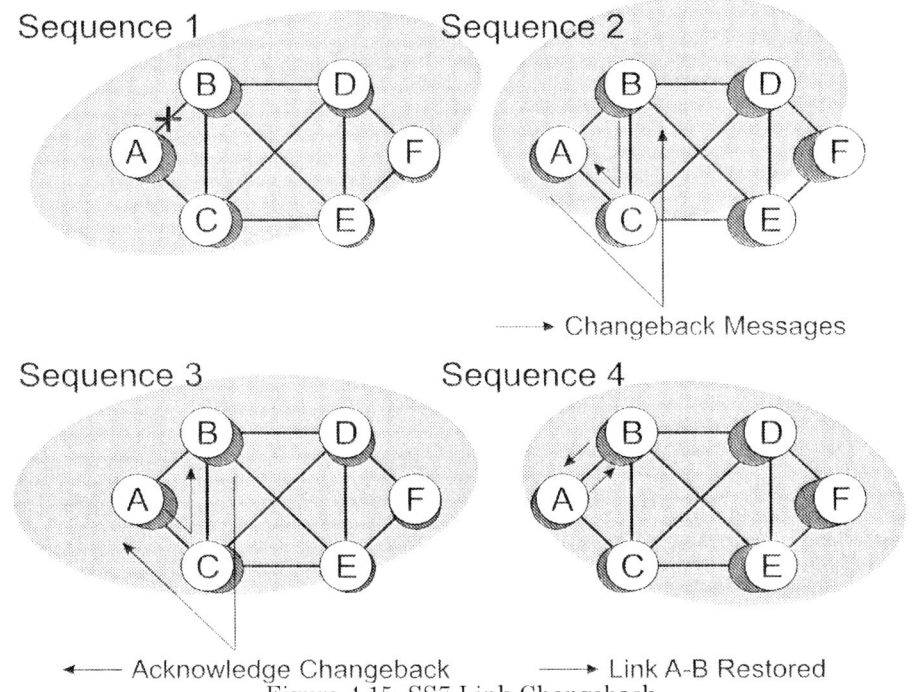

Figure 4.15. SS7 Link Changeback

Chapter 5

Signaling Connection Control Part (SCCP)

The Signaling Connection Control Part (SCCP) offers enhancements to the MTP Level 3 to provide connectionless and connection-oriented network services, as well as to address translation capabilities to the SS7 system.

Figure 5.1 shows how the SCCP functional part is located within the SS7 system. This diagram shows that the SCCP part provides enhancements to the MTP services and that the SCCP part is equivalent to the OSI Network model Layer 3.

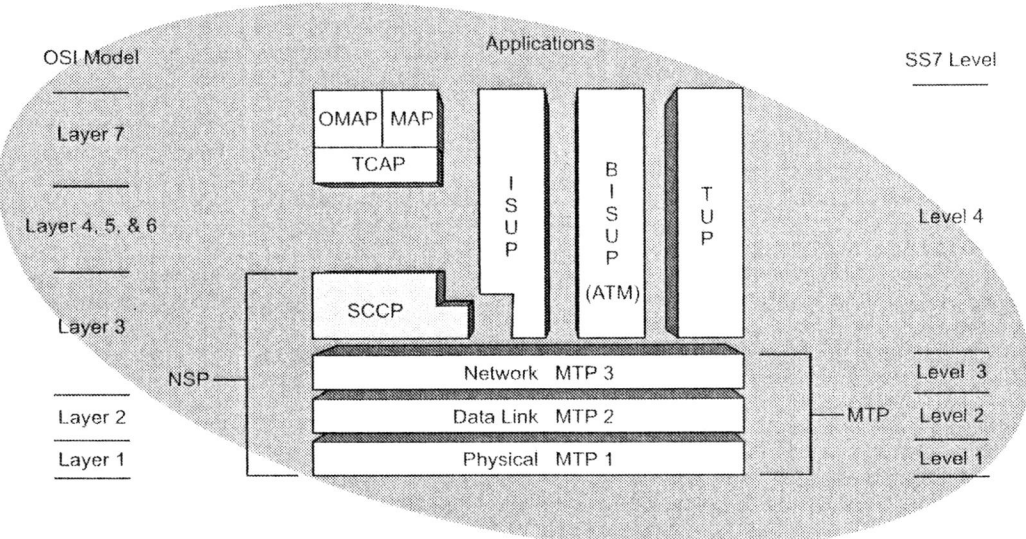

Figure 5.1, SCCP Function in the SS7 System Layers

SCCP Addressing Translation

The routing capability of the MTP is limited to delivering messages to the correct signaling point based on the destination point code (DPC). After receiving the message, it is forwarded to the correct MTP user within the signaling point based on the value of the service indicator contained within the signal information octet.

The SCCP provides an additional global title address translation function. A global title is an address, such as dialed digits for voice, data, ISDN or mobile networks, which cannot be routed on directly. The SCCP translates this number into a DPC and a sub-system number (SSN). The SSN identifies the SCCP user at a signaling point. Examples of SCCP users include SCCP management, ISDN-UP, and OMAP. The SSN is similar to the service indicator in the MTP routing but allows for 255 unique sub-systems to be defined at a signaling point, which the service indicator allows only 16 sub-systems to be defined.

Connectionless Service

The SCCP provides two classes of connectionless service: class 0 and class 1. In both classes, the SCCP accepts signaling messages from SCCP users and transfers them across the signaling network as independent messages unrelated to any previously sent messages.

The basic connectionless service (Class 0 service) does not provide for segmenting/reassembly, flow control or in-sequence delivery. The sequenced connectionless service (Class 1 service) is identical to Class 0 with one exception: it provides in-sequence delivery.

If the SCCP user requests in-sequence delivery of a stream of messages, the SCCP sets the signal link selection (SLS) code to the same value for all messages in the message stream. While this doesn't guarantee in-sequence delivery, the MTP routing and error recovery procedures provide high probability that the stream of messages will be delivered in sequence.

Figure 5.2 shows an example of steps that may be carried out for connectionless service:

> When an SCCP user requests transfer of information using connectionless service, the SCCP function at the local SSP (A) creates a unit data message containing the information. The SSP (A) transmits the unit data message to the remote SCCP contained in SSP (B). The information is then passed on to the SCCP user at that signaling point.
>
> Additional information may be transmitted as required. There is no connection establishment or release associated with the connectionless procedures.

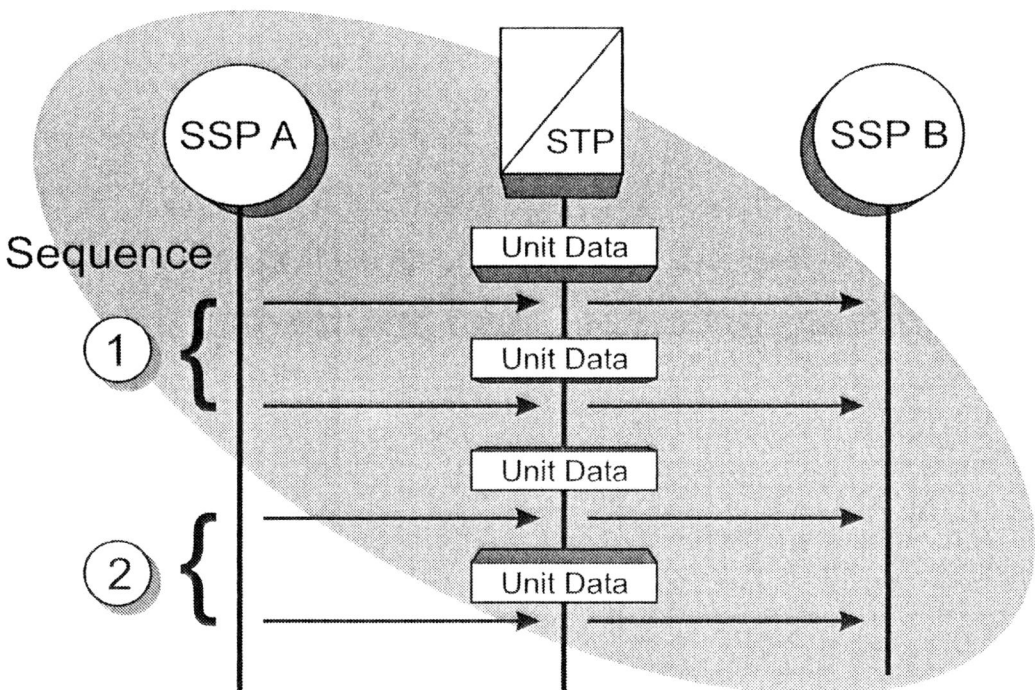

Figure 5.2. SS7 Connectionless Service

Connection Oriented Services

The SCCP provides two classes of connection-oriented services: class 2 and class 3. These services allow for the establishment of a temporary or permanent signaling connection to manage the transfer of messages between SCCP users. The signaling connection can be divided into three phases:

> Connection Establishment: In this phase signaling connection is established between two SCCPs.

> Data Transfer: Messages from SCCP users are exchanged across the signaling networks.

> Connection Release: The signaling connection between two SCCPs is disconnected.

The basic connection-oriented (Class 2) service provides a bi-directional transfer of messages between two SCCP users. The SLS field is set to the same value in all the messages to ensure in-sequence delivery. This class also provides segmenting and reassembly of SCCP user messages. If an SCCP user delivers a message to the originating SCCP that exceeds 255 bytes, the originating SCCP segments the message into more than one SCCP data transfer message. It then transmits those data transfer messages to the destination SCCP and the receiving SCCP then reassembles the original message for delivery to the destination SCCP user.

The flow control connection-oriented (Class 3) service provides flow control where all messages are assigned sequence numbers and the SCCPs monitor the data transfer to ensure in-sequence delivery. In the event of errors in the sequencing process or message loss, the signaling connection is reset and the SCCP users are notified of the event.

Figure 5.3 shows an example of steps that may be carried out for connection oriented service:

> The connection-oriented procedures are characterized by the originating and destination SCCPs setting up a logical signaling connection. When an SCCP user requests a connection-oriented SCCP service, the originating SCCP function at SSP (A) creates a connection request (CR) message and transmits it to the destination SCCP located in SSP (B). The CR contains the relevant setup information including a source local reference (SLR) number (SLR = 14). The SLR is used by SCCP (B) in all subsequent messages to identify this signaling connection.

> IF SCCP (B) determines that the called party is local user, and if resources are available, a connection confirmation (CC) is returned to SCCP (A). In the CC, SCCP (B) identifies the relevant signaling connection by referring to the original local reference (14) assigned by SCCP (A) in the CR. This is accomplished by setting the destination local reference (DLR) field to 14. SCCP (B) includes its own source local reference (SLR = 26). All subsequent messages related to this signaling connection include destination local reference numbers.

> Data messages can now be exchanged between the two end-points of the connection.

> When an SCCP user initiates a connection release, a released message is transmitted to the remote SCCP. When the released message is received at the remote SCCP, and indication is sent to thee SCCP user and a release complete message is sent back to the SCCP originating the release.

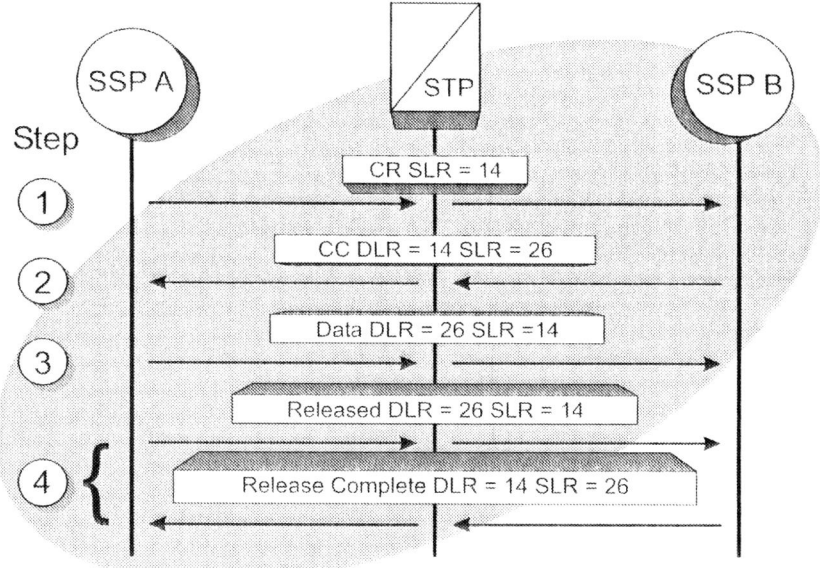

Figure 5.3. Connection-Oriented Service

SCCP Formats

SCCP messages are carried on the signaling data link within the Signaling Information Field (SIF) or Message Signal Units (MSU). The SCCP message is variable length and consists of an integral number of octets, or bytes.

Figure 5.4 shows that the SIF field of the MSU has a variable length to allow it to hold SCCP messages of different sizes.

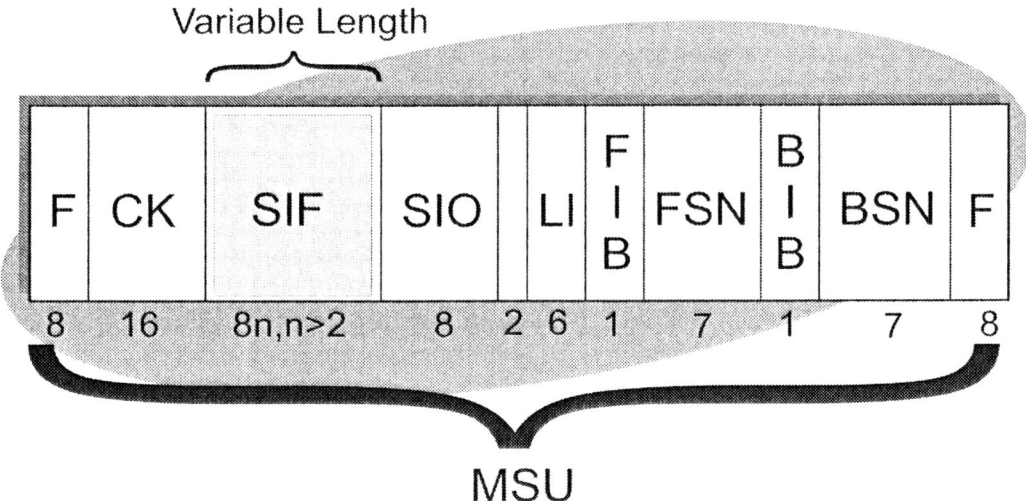

Figure 5.4, Message Signal Unit (MSU)

Figure 5.5 shows that the composed of a routing label (for delivery), message type (control), mandatory parts (required parameters), and optional information (optional parameters).

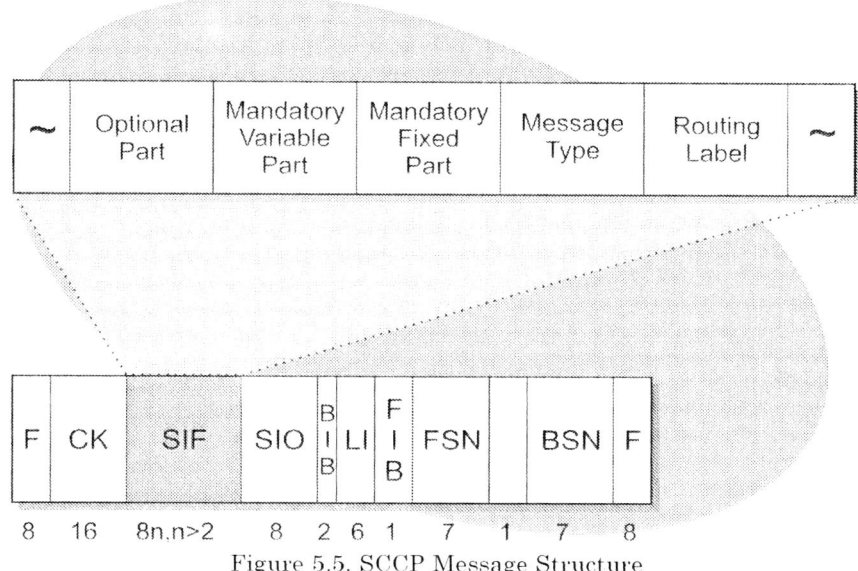

Figure 5.5. SCCP Message Structure

Routing Labels contain the information necessary for the MTP to route a message. The Message Type field is a one-octet field that uniquely identifies each message. Each SCCP message type has a defined format so that this field also identifies much of the structure of the remaining three parts of the message.

The Mandatory Fixed Part contains parameters that are both mandatory and have a fixed length for a specific message type. The message type defines the parameters that will follow, therefore the names and length indicators are excluded.

The Mandatory Variable Part contains parameters of variable length. Pointers are included in the message to indicate where each parameter begins. Each pointer is encoded as a single octet.

The Optional Part contains parameters that may or may not occur in any particular message type. Both fixed and variable length parameters may be included. A name and length indicator is included at the beginning of each optional parameter.

Chapter 6

Integrated Services Digital Network User Part (ISDN-UP)

The basic function of the Integrated Services Digital Network User Part (ISDN-UP) in the SS7 system is to control setup, connection, and release of circuit switched network connections between subscriber line exchange terminations. This includes basic voice and data services, and supplementary services.

Previous to developing the ISDN-UP, the standards groups developed the Telephone User Part (TUP) and the Data User Part (DUP). The ISDN-UP provides all the functions of both the TUP and the DUP, and is expected over time to replace their implementations.

Figure 6.1 shows how the ISDN-UP (ISUP) functional part is located within the SS7 system. This diagram shows that the ISDN-UP part provides the functionality of OSI Layers 3 through 7.

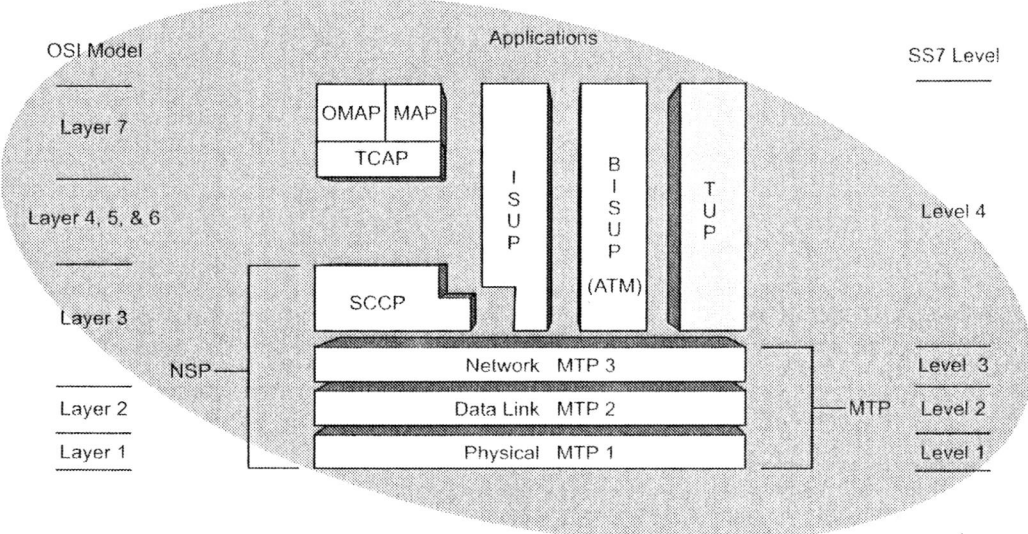

Figure 6.1. ISDN-UP (ISUP) in the SS7 Layers

Services

Services controlled by the SS7 system include voice and data transmission. Services can be divided into basic bearer services (e.g. voice transport) and supplementary services (e.g. call redirecting).

Basic Bearer Services

The basic bearer service for SS7 uses the ISDN-UP to control of 56 Kb/s or 64 Kb/s circuit-switched voice or data connections. This service, termed the basic bearer service is divided into three phases (in this way, it is similar to the SCCP connection-oriented service): call setup, connection, and call release.

Call Setup

Call setup involves the initiation of a call, determination of the destination of the receiving device (e.g. destination switching exchange), alerting of the destination device that a call is to be received, and connection of a voice path through the network.

Figure 6.2 shows an example of how the call setup process can be performed in an SS7 system:

The ISDN-UP call setup procedures begin when the calling party initiates a call using the applicable access signaling. In this example, the calling party transmits an ISDN "setup" message.

When the origination exchange has received the complete selection information from the calling party, and has determined that the call is to be rerouted to another exchange, selection of a suitable, free, inter-exchange circuit takes place. An Initial Address Message (A\IAM) containing information requ9itred to reroute the call to the destination exchange is sent by each exchange until the call reaches the destination exchange.

The destination exchange, on receiving the IAM, notifies the called party of the incoming call using the appropriate access signal.

The called party normally responds with an alerting indication that is passed backwards through the network as the Address Complete Message (ACM). When the originating exchange receives the ACM, an alerting message is passed to the calling party using the applicable access signal. At this point the caller hears the ringback tone.

When the called party answers the call, a connection message is returned to the destination exchange, an answer message (ANM) is then passed backwards through the network. Call charges normally begin when the ANM is returned to the originating exchange.

When call setup is complete a connect message is returned to the calling party.

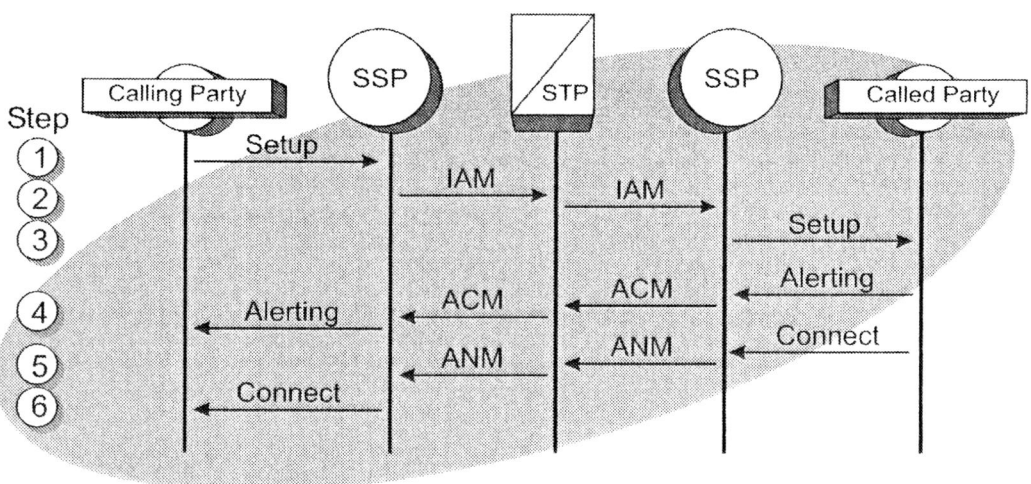

Figure 6.2. Call Setup

Connection

To setup and maintain a connection, messages are sent end-to-end that contain information that is required by the end-points of a circuit-switched connection. These end-points may be local exchanges or international gateway exchanges.

Figure 6.3 shows that end-to-end messages may be transmitted during the call setup or connection phase of a call through a SS7 system. This example shows that the SS7 messages can originate at either switch (SSP) and that these messages are routed to their destination switch (SSP) through switching transfer points (STPs).

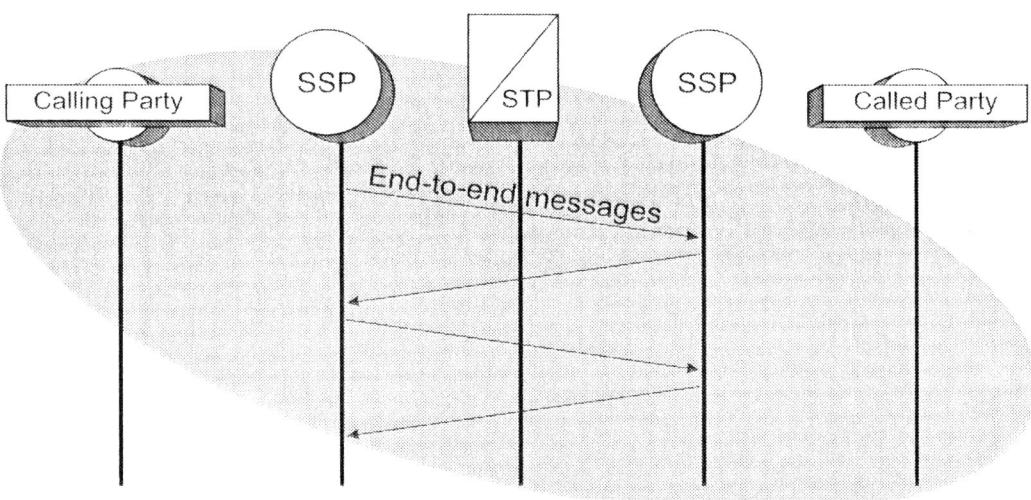

Figure 6.3, SS7 Connection Phase

Call Release

To release (disconnect) a connection, a call release procedures is initiated when either end of a call sends a disconnect signal. This call release procedure may occur when either user terminates the call (e.g. hangs-up the telephone) or when a connection path between the users becomes unavailable (e.g. loss of connection.)

Figure 6.4 shows example of how a call release process may occur in an SS7 system:

In the example shown, the calling party has initiated the call release, by sending a disconnect message to the originating exchange.

The originating exchange then sends a release message to the tandem exchange and returns a release message to the calling party.

On receiving the release message, the tandem exchange returns a release complete (RLC) message back to the originating exchange and forwards the release message on to the destination exchange.

When the destination exchange receives a release message, it forwards a disconnect message to the called party and returns an RLC message back to the tandem exchange.

On receipt of a disconnect message from the destination exchange, the called party returns a release message back to the destination exchange.

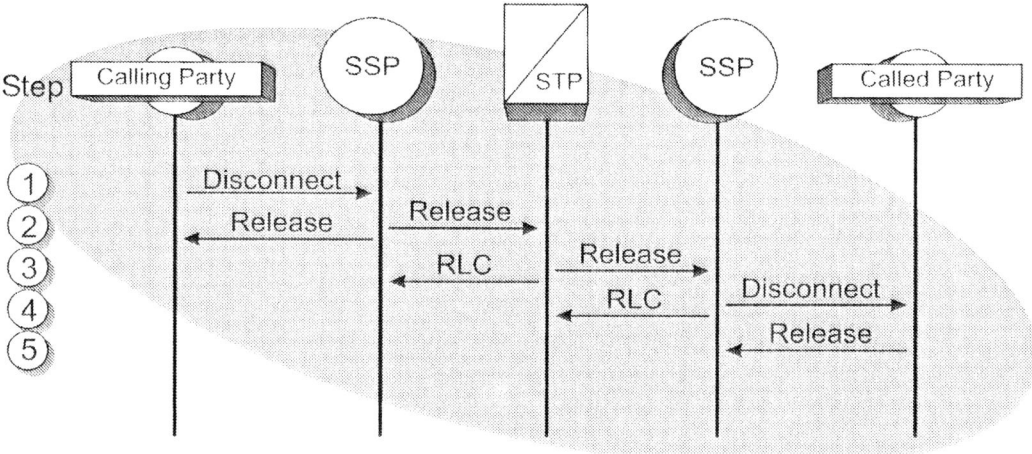

Figure 6.4. SS7 Call Release

Supplementary Services

Supplementary services provide a network user with capabilities beyond those of elementary call control. Supplementary services enrich the basic services functions and are not specific to a telephone or system features. Often, the subscriber (user) can specify some of the operations of supplementary services (such as call redirection). Some of the supplementary services for the SS7 system include redirection of calls, malicious caller identification, called and calling line identification, closed user group (CUG), and call completion to a busy subscriber.

Redirection of Calls

This facility enables a user to have calls redirected to another predetermined number during periods when the facility is activated.

Figure 6.5 shows an example of how redirection of calls may be accomplished in an SS7 system:

In this example, user A activates call redirection to redirect calls to User B.

User C calls User A. The call is redirected to User B.

The SS7 signaling network completes the call directly between User C and User B without tying up circuits at exchange A. This is the major advantage of call redirection over call forwarding. In the case of call forwarding, the call from User C would be routed to the switch serving User A and then tie up a second circuit when leaving switch A on its way to user B.

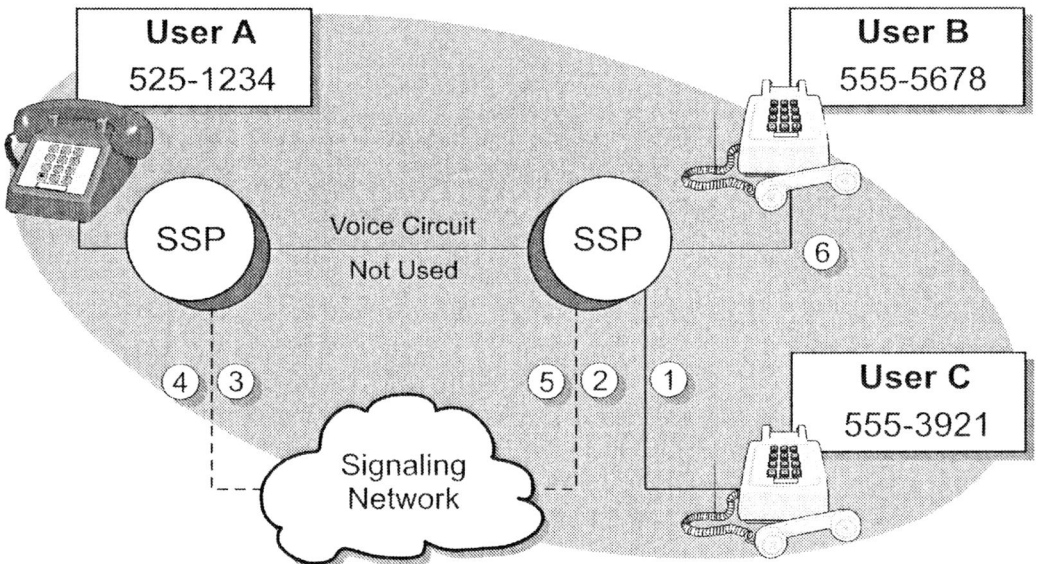

Figure 6.5. SS7 Redirection of Calls Facility

The Redirection of Calls Information Prohibited facility enables the user who has activated the redirection of calls facility to prevent the calling party from being informed that the call is redirected.

Malicious Caller Identification

This facility is a user-initiated request for the identification of the calling party and the called party.

Figure 6.6 shows the steps that may be used for malicious caller identification in an SS7 system:

Malicious caller places a call from station B to user A.

User A initiates call trace once the call is recognized as a malicious call.

ISDN-UP sends the calling line ID of the malicious caller, called line ID of user A, and time and date of the call to the designated central location.

The information is stored for future investigation.

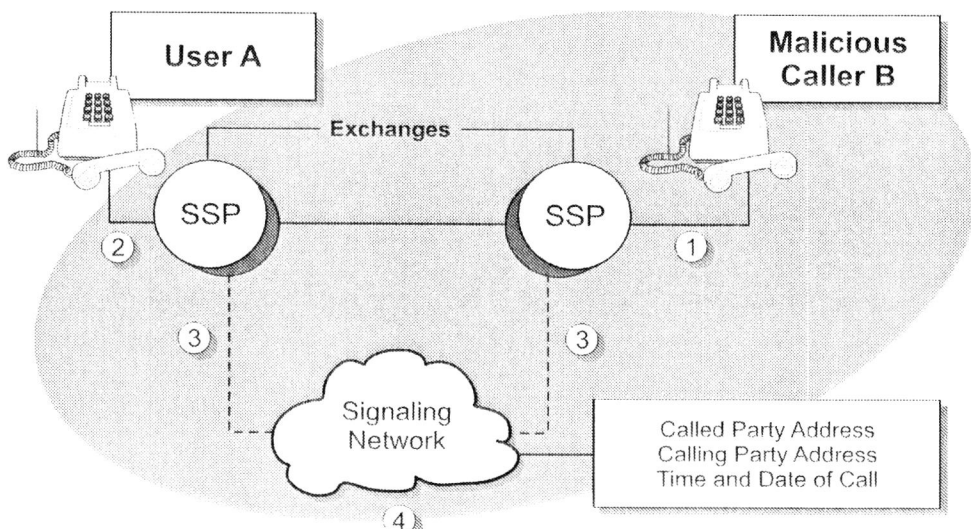

Figure 6.6. Malicious Call Identification

Calling Line Identification

With this service, the calling line ID is presented to the called party during call setup. This service is at the center of a "right-to-privacy" controversy where a number of groups are challenging the telephone companies' right to offer this service. In response to this controversy, some telephone companies are offering subscribers the option to restrict presentation of their telephone number to the called party.

Called Line Identification

This service provides the calling party with the identity of the user to which the call has been connected. In the case where the call has been forwarded, the call forward reason is also displayed. The reason that the call was forwarded may be displayed for the caller's information; reasons for the call being forwarded may be that the line is busy, there is no answer, or that all calls are being forwarded.

Figure 6.7 shows an example of how called line identification may be performed in an SS7 system. In this example, user A would have user B's number (555-5678) showing on his display. If the called party has the address presentation restriction facility active, the called party address would not be presented to the calling party.

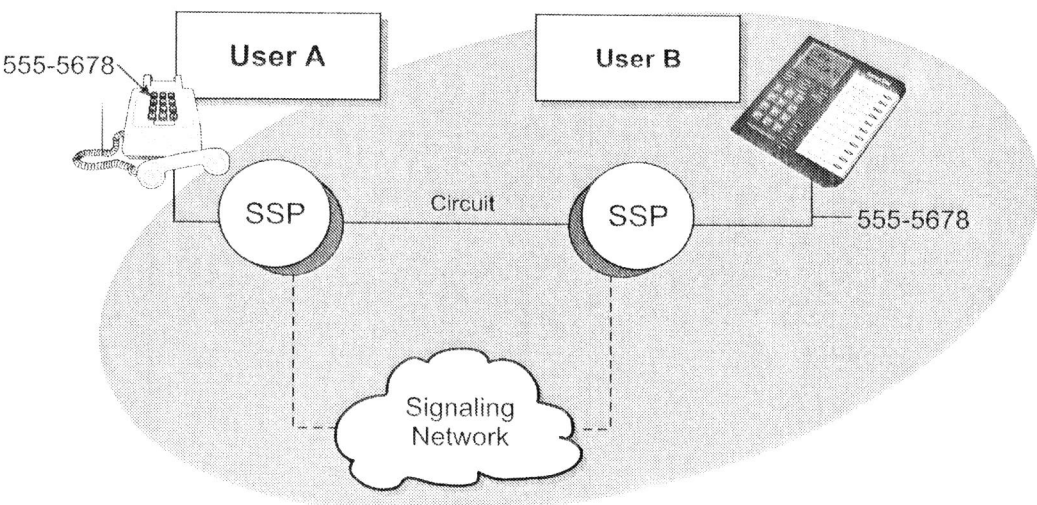

Figure 6.7. SS7 Called Line Identification

Closed User Group (CUG)

A Closed User Group (CUG) facility enables users to from groups with different combinations of restrictions for access to or from users having one or more of these facilities. The following CUG facilities are standardized:

Closed user group with intercommunication only among members of the CUG;

Closed user group with outgoing access;

Closed user group with incoming access;

Incoming calls barred within the closed user group; and,

Outgoing calls barred within the closed user group.

A user may belong to one or more CUGs. When this is the case, one CUG is nominated as the user's preferred CUG. Different combinations of CUG facilities may apply for different users belonging to the same CUG.

Completion of Calls to Busy Subscriber

The Completion of Calls to Busy Subscriber (CCBS) facility enables a calling party encountering the busy condition to complete the call automatically when the called party becomes free (e.g. without redialing.) This facility is also known as network "ring again."

The calling party activates the user facility by making a request to the exchange to which he is connected. When the facility is activated, the status of the called party address is continually tested by its local exchange. When the address is free, the calling party is alerted. When the calling party answers, the call attempt is made once again.

ISDN-UP Message Formats

The ISDN-UP information is carried in the Signaling Information Field (SIF) of the Message Signal Unit (MSU). The ISDN-UP message is variable length but consists of an integral number of octets.

Figure 6.8 shows that the MSU has a variable length SIF field to allow the different lengths for ISDN-UP messages.

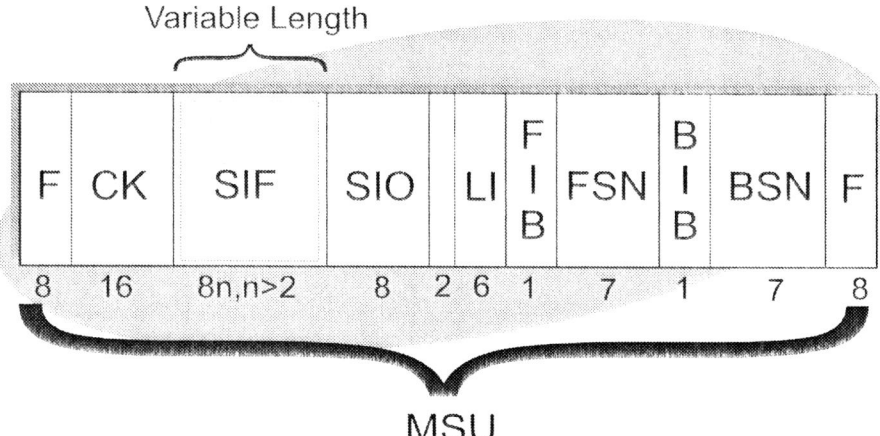

Figure 6.8. Variable Length Message Signal Unit (MSU) Structure

Figure 6.9 shows that the ISDN-UP message is composed of a routing label (for call setup and delivery) connection identification code (for circuit identification), message type (control), mandatory parts (required parameters), and optional information (optional parameters).

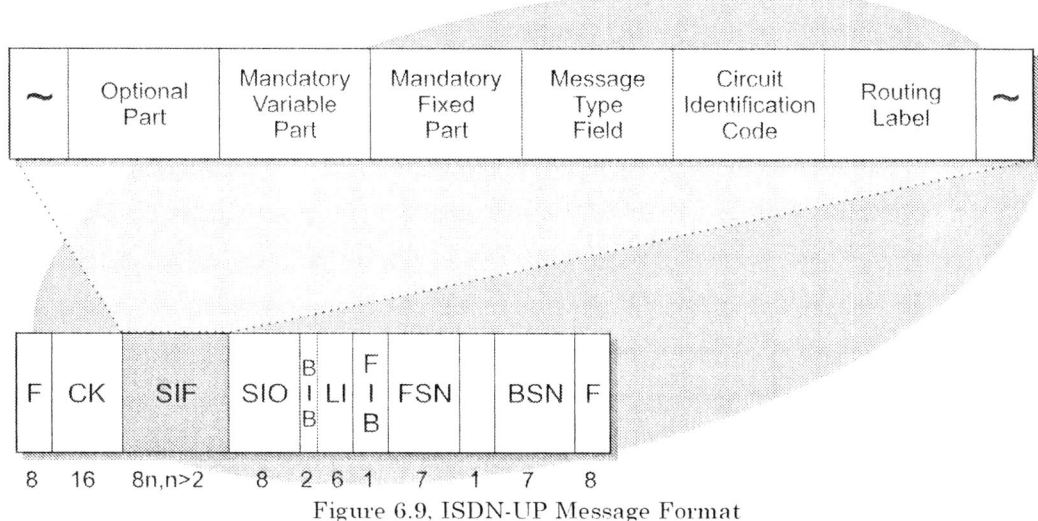

Figure 6.9, ISDN-UP Message Format

Routing Label

This label contains the information necessary for the MTP to route the message. The routing label is examined by the message transfer part (MTP) to determine if additional routing is required or if the message is designated for it's on SS7 node.

Circuit Identification Code (CIC)

This code identifies the circuit to be assigned to a call. For circuits that are derived from a 2048 Kb/s digital path, the Circuit Identification Code contains the five least significant bits, the binary coding of the time slot assigned to the speech circuit. For circuits that are derived from an 8448 Kb/s digital path, the circuit identification code contains the seven least significant bits, the binary coding of the time slot assigned to the speech circuit. In either case, when necessary, the remaining bits are used to identify one among several systems interconnecting an originating and destination point. For circuits derived from other digital paths (for instance DS-1s in North America,) similar CIC assignments are used.

Figure 6.10 shows that the CIC field is located in the SIF field of the MSU. The CIC field is variable length and it divided into most significant bits, least significant bits, and spare bits.

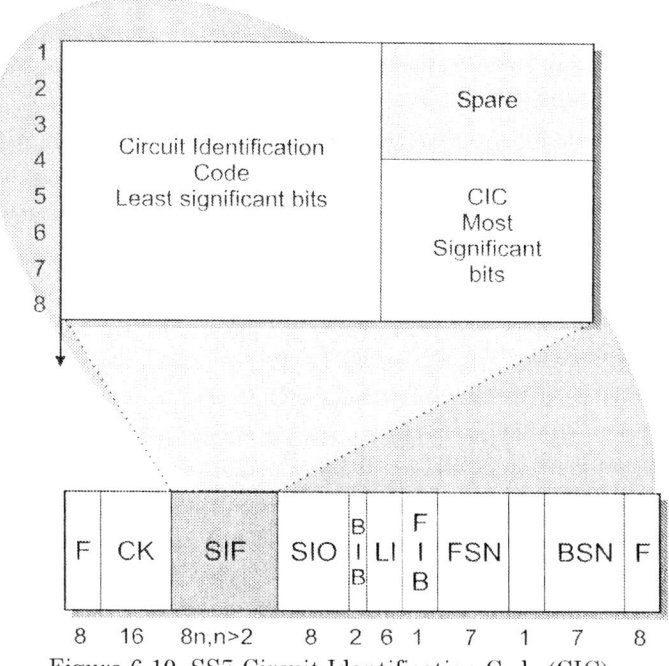

Figure 6.10. SS7 Circuit Identification Code (CIC)

Message Type Field

This field is a one-byte field identifying the message, for instance, Initial Address Message, Release Message. Each ISDN-UP message type has a defined format so that this field also identifies much of the structure of the remaining parts of the message.

Mandatory Fixed Part

Those parameters that are mandatory and of that have a fixed length for a particular message type are contained in the mandatory fixed part. The position, length, and order of the parameters in the fixed part are uniquely defined by the message type field. As a result, the names of the parameters and the length indicators are not included in the message.

Mandatory Variable Part

Some of the ISDN-UP messages will include parameters that have variable length that are associated with mandatory fields. Mandatory parameters of variable length are included in the variable length mandatory part. Pointers are used to indicate the beginning of each parameter. Each pointer is encoded as a single octet. The name of each parameter and the order in which the pointers are set are implicit in the message type.

Optional Part

This part consists of parameters that may or may not occur in any particular message type. Both fixed length and variable length parameters may be included. Optional parameters may be transmitted in any order. Each optional parameter includes the parameter name (one octet) and the length indicator (one octet) followed by the parameter contents.

Chapter 7

Transaction Capabilities Application Part (TCAP)

Introduction

The Transaction Capabilities Application Part (TCAP) is an SS7 application protocol that is used by a variety of distributed applications. TCAP provides non-circuit related information transfer capabilities and generic services to applications, yet TCAP remains independent of the application.

Figure 7.1 shows how the TCAP functional part is located within the SS7 system. This diagram shows that the TCAP part converts and manages messages between lower layers and the application services and when TCAP is used for mobile services it is called mobile application part (MAP). This diagram also shows that the TCAP functional part in the SS7 system is equivalent to the OSI network model application layer 7.

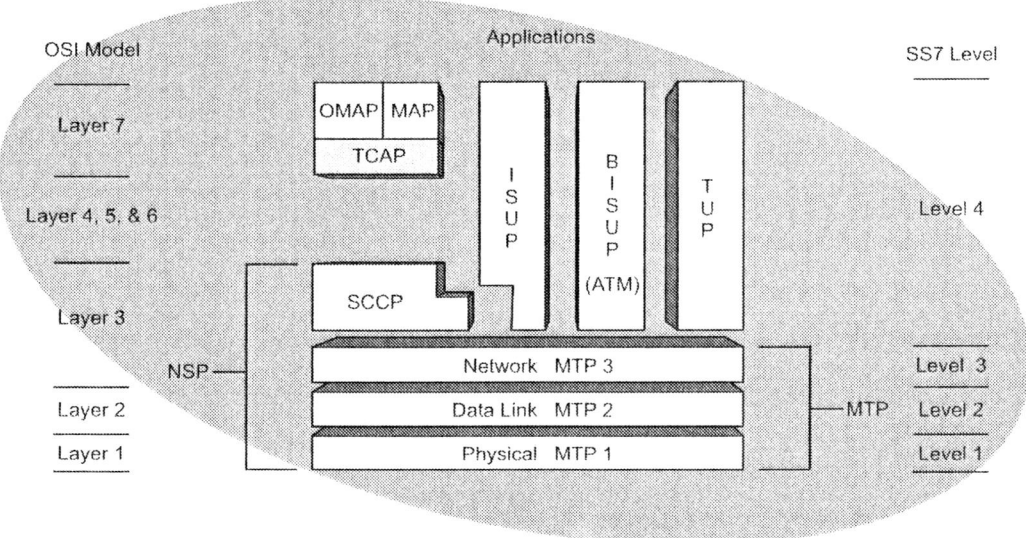

Figure, 7.1, TCAP in the SS7 Layers

Application process services (such as 800 or Freephone services or networking again) use TCAP to provide enhanced network services and operations, administration and maintenance (OAM) functions. An application process requiring service from TCAP is called a Transaction Capabilities User or TC-User. TCAP services may be used between:

> Signaling points;
> Signaling points and network service centers (such as data bases or an OAM center),
> Network service centers.

TCAP itself does not provide any services to telecommunications network users. Instead, it provides the capability for a large variety of distributed applications to invoke procedures at remote locations on the SS7 network. A common procedure is the query of a Service Control Point (SCP) database.

For example, 800 or Freephone service uses the TCAP protocol to pass the dialed 800 numbers to an SCP database and request a translation to a routing number. The routing number is then returned to the signaling point to allow call routing.

TCAP services are based on a connectionless network service. As seen in Figure 7.1 and Figure 7.3, currently no services are provided from the session, presentation or transport layers. TCAP interfaces directly with the SCCP, making use of the SCCP connectionless service to transfer between two TCAPs applications.

TCAP Message Formats

TCAP messages are contained within SCCP message signal units. The TCAP message format consists of two parts, the transaction sub layer and the component sub-layer.

Figure 7.2 shows that the TCAP Message Signal Unit follows an SCCP message header and that the SCCP message is located in the SIF field of the MSU.

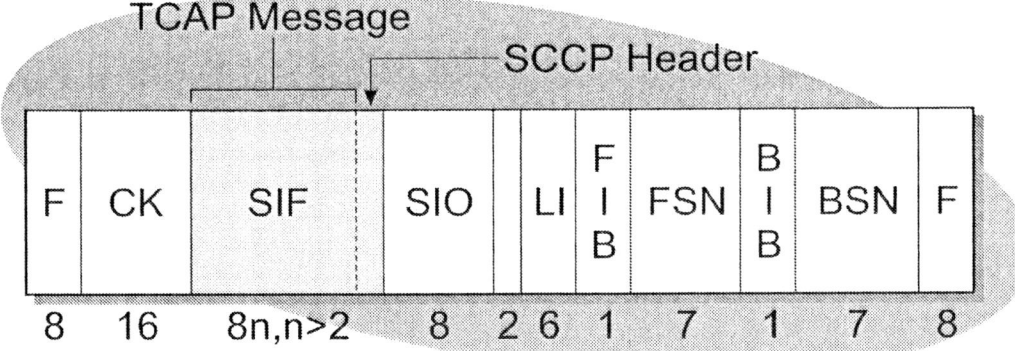

Figure 7.2. TCAP Message Signal Unit Inside SCCP Message

Figure 7.3 shows that TCAP message is divided into two of three parts that support the application layer communication requirements. The third part of the application communication support includes the TC-User part.

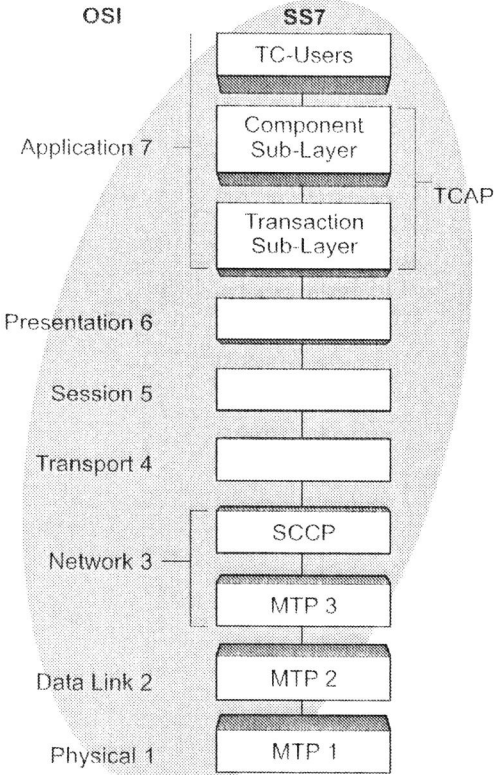

Figure. 7.3. TCAP Sub-Layers

The component sub-layer is responsible for accepting components from TC-Users and delivering those components in order to the remote TC-User. The transaction sub-layer is responsible for managing the exchange of messages containing components between two TCAP entities. This exchange of components, to perform an application, is called a dialogue.

TCAP Component Layer Types

There are five types of components that may be present in the component portion of a TC message. These are invoke component,

Invoke Component

This component requests that an operation be performed. IT may be linked to another invocation previously sent by the other end.

Return Result (not last) Component

When TCAP uses a connectionless network service, it may be necessary for the TCAP user to segment the result of an operation. It this case this component is used to convey each segment of the result except the last, which is conveyed in a Return Result (last) Component.

Return Result (last) Component

This component reports successful completion of an operation. It may contain the last/only segment of a result.

Return Error Component

This component reports that an operation has not been successfully completed.

Reject Component

This component reports the receipt and rejection of an incorrect component, other than a Reject Component.

Transaction Sub-layer

There are five transaction sub-layer message types defined; begin, continue, end, unidirectional, and abort messages. Begin, continue, and end messages are used in structured dialogues, while unidirectional are for unstructured dialogues, and abort is reserved for abnormal situation.

Figure 7.4 shows the Transaction Sub-layer messages types and flows. In order of transmission, the message types are: Begin, Continue, End, Unidirectional, and Abort.

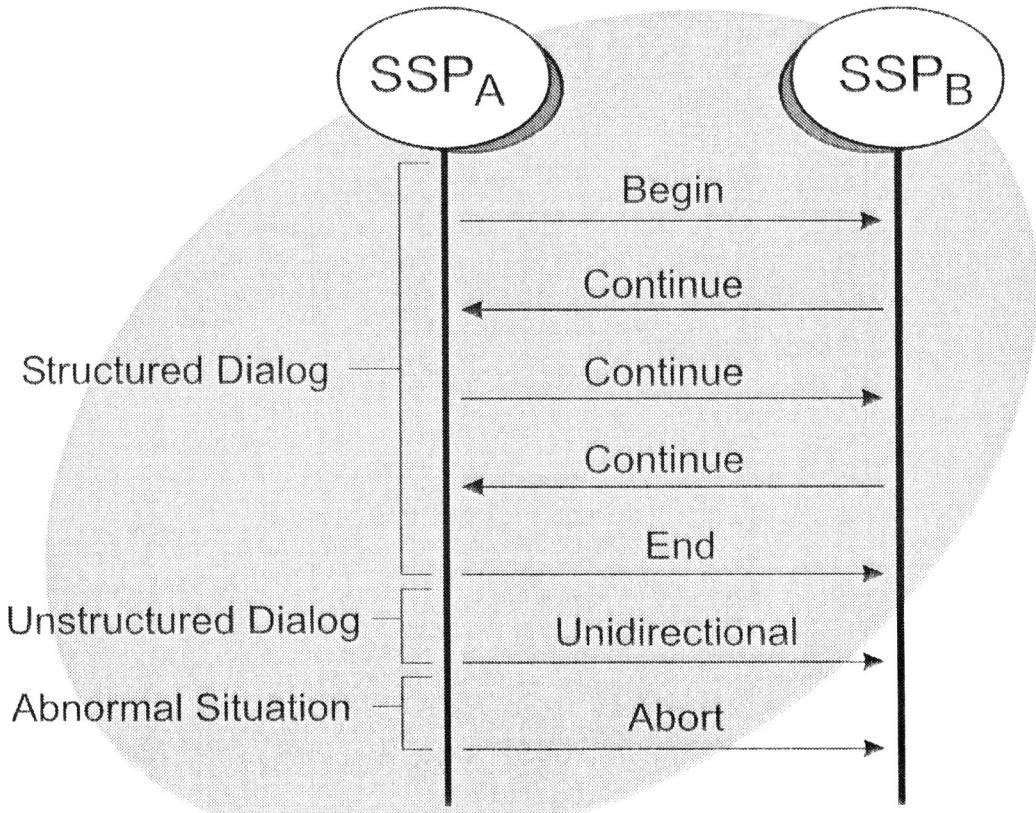

Figure 7.4. Transaction Sub-Layer Message Flow

Begin

The Begin message is used to establish a transaction with a remote peer transaction sub-layer. The Begin message may include one or more components.

Continue

The Continue message is used following a Begin message to transfer additional information related to the transaction. The Continue message contains one or more components.

End

The End message is used to terminate a transaction. All remaining untransmitted components are included in the End message.

Unidirectional

Unidirectional messages are associated with unstructured dialogue; such a dialogue occurs when a TC-user needs to send one or more components to a remote TC-user and replies are not expected.

Abort

The Abort message is used to terminate a transaction following an abnormal situation detected by a transaction sub-layer or a request by the component sub-layer to abort the transaction.

Information Elements

The transaction and component sub-layers are structured in what are called information elements. An information element consists of three fields; a tag octet (whereby an octet equals 8 bits), a length octet, and a variable length contents field. The tag identifies the information element; the length specifies the number of octets in the information element, excluding the tag or length octets; and the contents contains the information that the element is intended to convey. There are two structures that an information element may have: a primitive form, and constructor form.

Figure 7.5 shows the different types of SS7 TCAP information element forms. A primitive form is one whose contents field does not contain additional information elements. A constructor form is one where the contents field contains one or more information elements. These information elements may in turn themselves be constructor form.

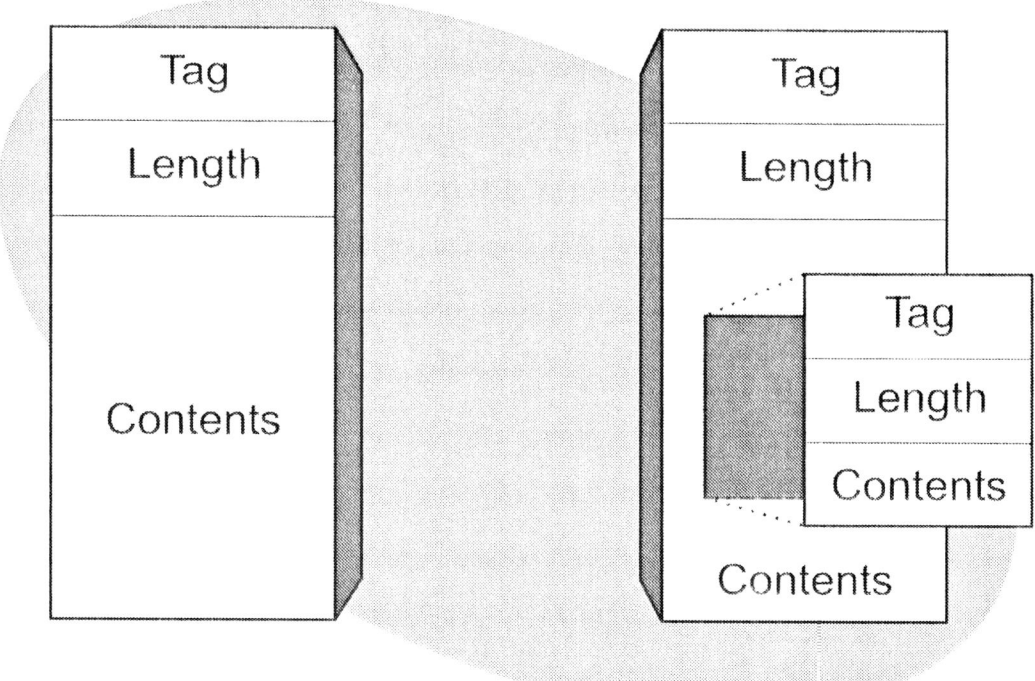

Figure 7.5. Information Element Forms

TCAP Applications

TCAP makes possible a wide variety of advanced network services based on information exchange between network components. These applications include network ring again, calling/credit card validation, and 800 Freephone Service.

Network Ring Again

Network ring again can be invoked after a caller from one switch encounters a busy signal when calling someone on another switch. This feature allows the call to be re-established when the called party hangs up.

Figure 7.6 shows and example of how the network ring again application may be performed in an SS7 system:

> User A places a call to User B. IAMs are sent to the destination exchange. Call Setup was described in detail in Chapter Six.
>
> A Release (REL) message is returned to User A indicating that user B is busy. RLCs are returned to complete the release sequence.
>
> User A may request Network Ring Again in which case a TCAP message requesting ring again is sent to exchange B.
>
> User A's Ring Again request is acknowledged by exchange B.

Exchange B monitors the busy/idle status of User B's line. When User B hangs up, exchange B returns a TCAP station idle message back to exchange A.

Exchange A returns a TCAP enc message to complete the TCAP dialog.

Exchange A alters User A with a distinctive ring, and if User A picks up the phone, exchange A proceeds to re-establish the call to User B.

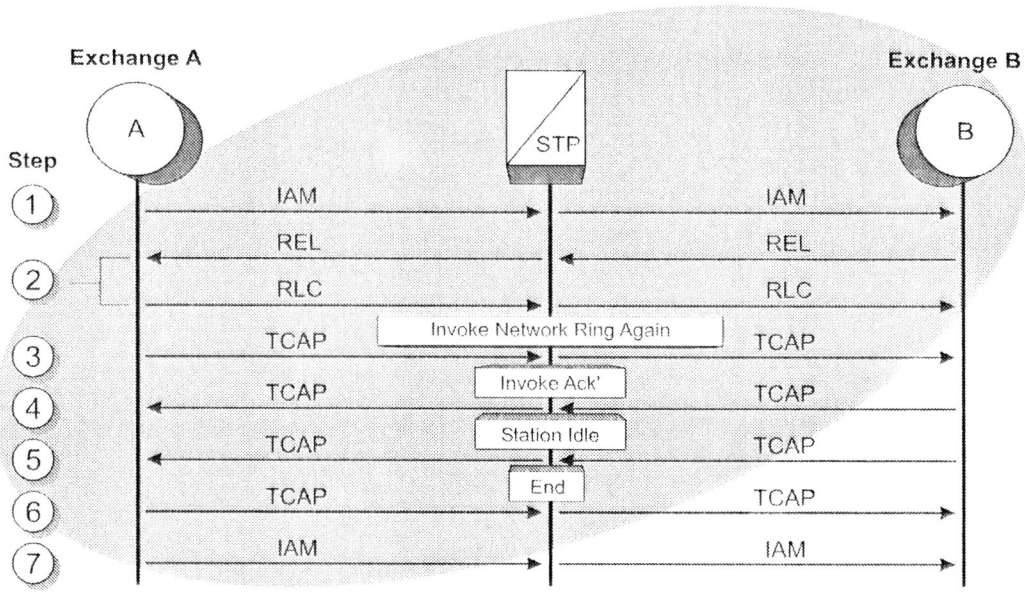

Figure 7.6. SS7 Network Ring Again

Credit/Calling Card Verification

Credit/Calling Card Verification allows callers away from home or the office to place toll calls with the convenience of billing the call to a credit/calling card.

Figure 7.7 shows an example of the steps that may be performed to provide for credit/calling card applications in an SS7 system:

A caller dials a number with a request to bill the call to a credit/calling card.

Exchange A sends a request to an SCP containing the credit/calling card database to verify that he card id valid.

The SCP credit/calling card database verifies the card number.

A reply back to exchange A indicates whether or not the card number is valid.

If valid, Exchange A proceeds to place the call.

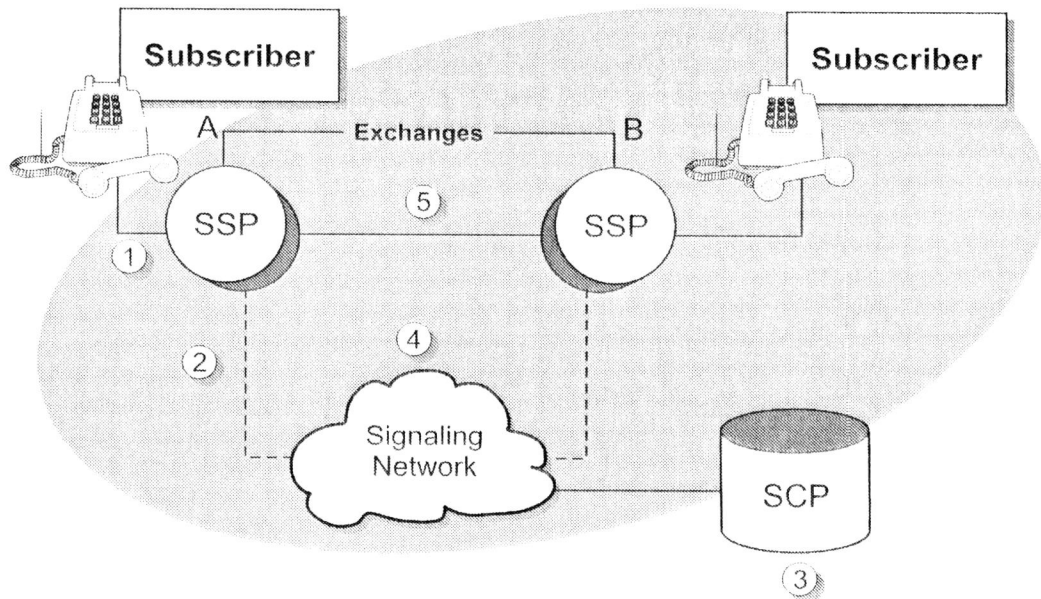

Figure 7.7. SS7 Credit/Calling Card Processing Example

800 Freephone Service

800 Freephone service that allows callers to dial a telephone number without being charged for the call. The toll free call is billed to the receiver of the call. In the Americas and other parts of the world, Freephone is sometimes called "Toll Free" and they typically begin with 800, 888, 877, or 866. Enhanced 800 services allow telecommunications carriers to offer a number of flexible 800 services tailored to customer requirements.

Based on key parameters (e.g. the time of day or the location of the caller,) 800 calls to a specific 800 number could be routed to different locations. This arrangement would allow multi-location businesses or governments to tailor the services they provide to their customers.

Figure 7.8 shows an example of how the SS7 system may provide toll free/freephone (e.g. 800 number) services:

A caller at exchange A dials an 800 number.

Route to SCP for Translation.

The 800 database located at the SCP determines the routing information for the call.

The SSP Route to Actual Number.

Exchange A then proceeds to set up the call.

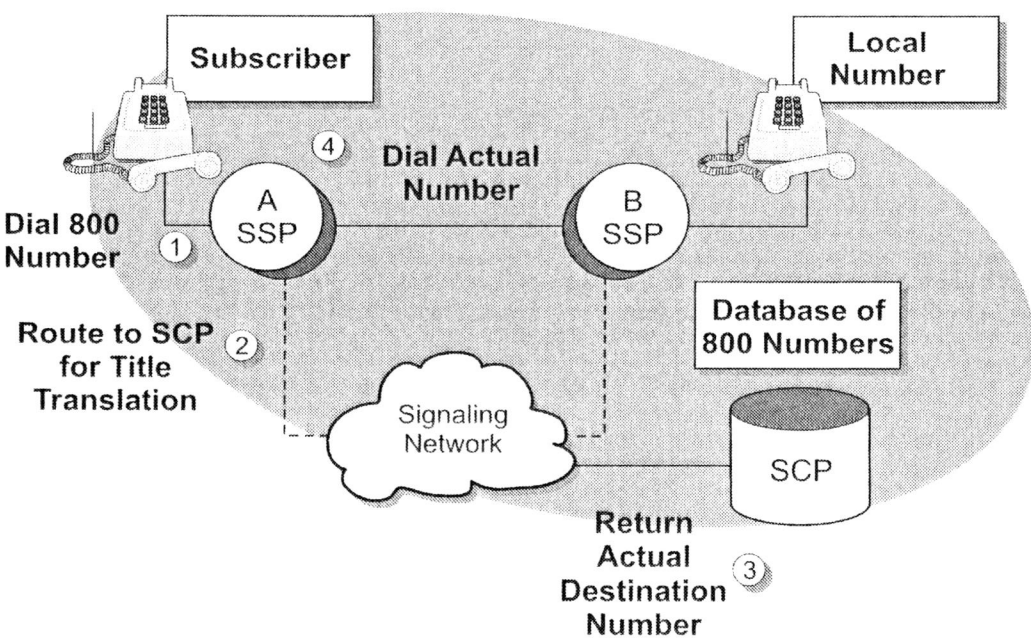

Figure 7.8, SS7 Toll Free/Freephone (800 number) Service Processing

Chapter 8

Operations, Maintenance and Administration Part (OMAP)

The Operations, Maintenance and Administration Part (OMAP) provide the procedures that are used to operate, maintain and administer the functions that are associated with the SS7 network.

Figure 8.1 shows how the OMAP functional part is located within the SS7 system. This diagram shows that the OMAP functional part uses the TCAP function to communicate with the SS7 system and that the OMAP functional part in the SS7 system is equivalent to the OSI network model application layer 7.

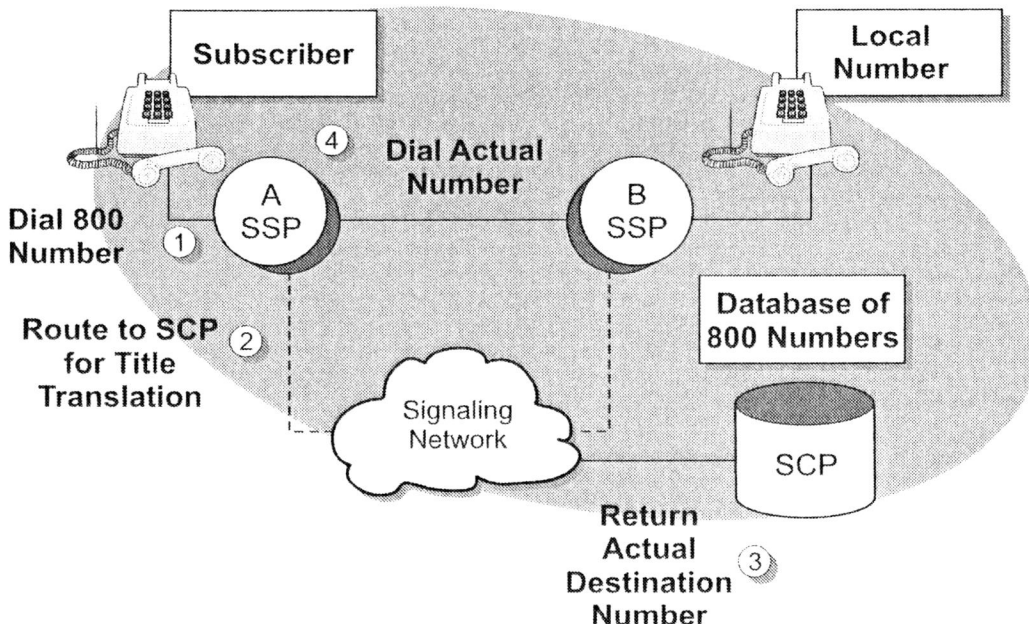

Figure 8.1. OMAP in the SS7 Protocol

Management Model

The management model coordinates the resources in an SS7 system. The network management model allows an operation center terminal (computer) to monitor and control functions and services to remote locations throughout the SS7 system.

Figure 8.2 shows the management model for the SS7 system. This example shows that the System Management Application Process (SMAP) application coordinates the system through the system management application entity (SMAE) protocol layer and that the SMAE layers use the Layer Management Entity (LME) to communicate using the Layer Management Interface (LMI) format that can communicate and store information in a management information base (MIB).

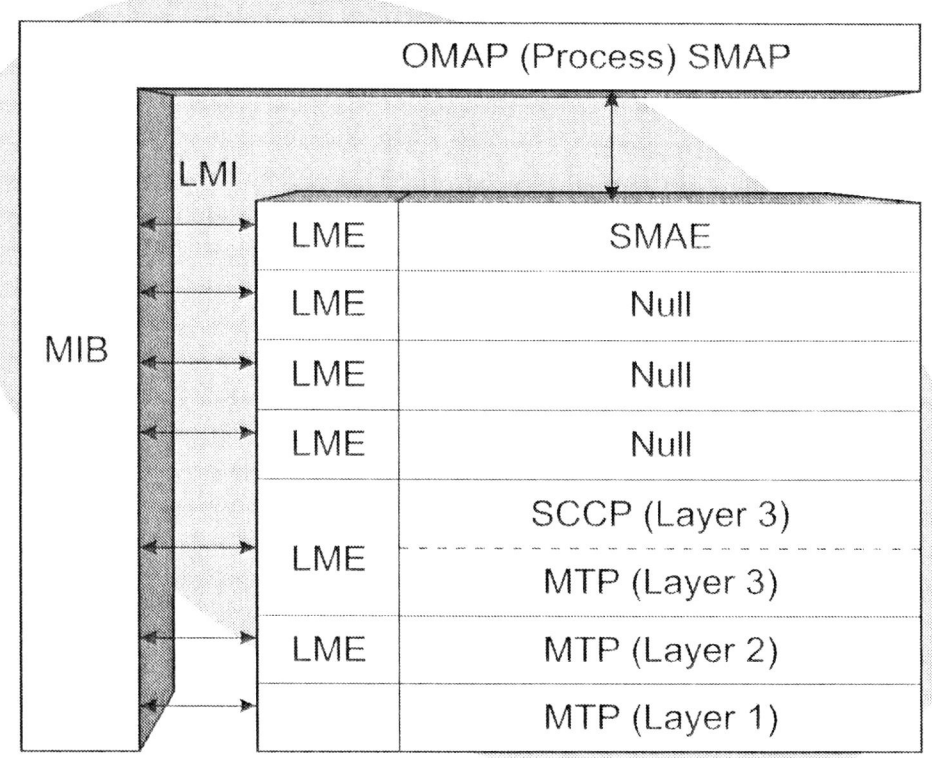

Figure 8.2. SS7 Management Model

System Management Application Process (SMAP)

The System Management Application Process (SMAP) monitors, controls and coordinates resources through the Application Layer protocols. The SMAP uses SMAE protocol layer to communicate with network equipment and systems.

Management Information Base

The Management Information Base (MIB) contains the performance and alarm data collected from the SS7 network by OMAP and the configuration data accessed by OMAP to assist in the management of the SS7 network.

The information transfer between equipment and systems is performed by the Layer Management Interface (LMI). The LMI is implementation-dependent and not subject to standardization. Data collected from the protocol layers is stored in the MIB using the LMI. The Layer Management Entity (LME) performs the management functions when communicating with it's corresponding SS7 layer.

System Management Application Entity (SMAE)

The System Management Application Entity (SMAM) is the aspect of SMAP involved with communications. SMAE contains one or more communications functions for an application. Each communications function is called an Application Service Element (ASE).

OMAP Application Service Elements (ASE)

Two OMAP application service elements (ASE) are defined: The Message Routing Verification Test, and the Circuit Validation Test. The OMAP ASEs use the services of TCAP to perform their functions.

Message Routing Verification Test (MRVT)

The Message Routing Verification Test (MRVT) is designed to verify that the routing tables in SS7 network nodes are consistent and that no routing loops or other anomalies are present.

Figure 8.3 shows an example how Message Routing Validation Test (MRVT) may be performed in an SS7 system:

At the initiating signaling point (A), a Message Route Verification Test (MRVT) message is sent on each signaling route which is contained in the MTP routing table for the test destination; The initiating signaling point then waits for Message Route Verification Acknowledgement (MRVA) and Message Route Verification Result (MRVR) messages to be returned.

At intermediate signaling points, if the test can be run and the initiating and test destination signaling points are known, an MRVT message is sent on each signaling route which is contained in the MTP routing table for the test destination.

Again, at intermediate signaling points, if the test can be run and the initiating and test destination signaling points are known, an MRVT message is sent on each signaling route which is contained in the MTP routing table for the test destination.

When the test destination receives an MRVT message, it checks the MRVT message to see if the initiator of the test is known and if a trace has been requested. If the initiator is known, this part of the overall test is considered successful. An MRVA message is returned to the signaling point(s) that sent the MRVT message(s) to the test destination. IF a trace has been requested, the test destination also sends a MRVR message directly to the initiator of the test.

At the intermediate points, the MRVA messages received in step 4 are in response to previously sent MRVT messages from step 3. The information contained in the received MRVA messages is forwarded to the signaling point(s) that originally sent MRVT messages to the intermediate points. Under certain circumstances messages to the intermediate points. Under certain circumstances an MRVR message may be sent directly to the initiator of the test.

Again, at intermediate signaling points, the MRVA messages received are in response to the previously sent MRVT messages. The information contained in the received MRVA messages is forwarded to the signaling point(s) that originally sent MRVT messages to the intermediate point. Under certain circumstances an MRVR message may be sent directly to the initiator of the test.

The MRVT is now complete; test results are forwarded to the OMAP process requesting the test.

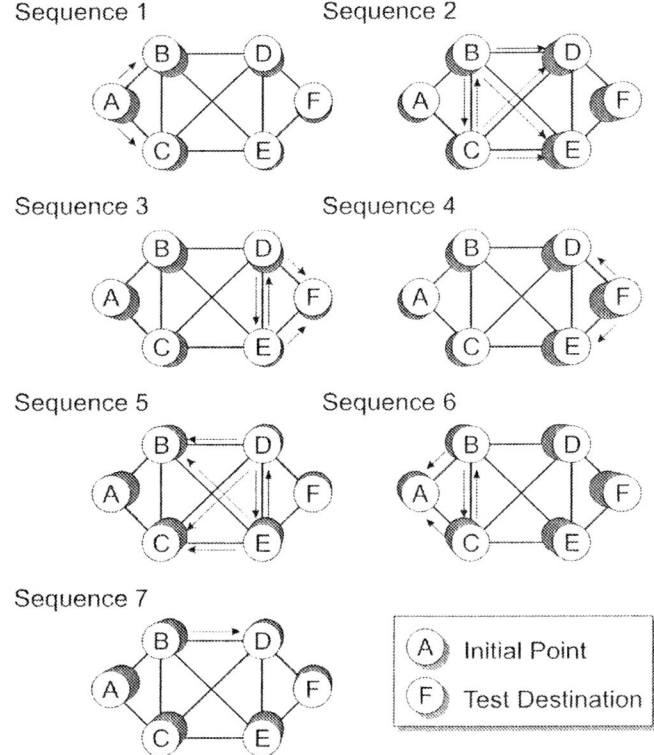

Figure 8.3, SS7 Message Routing Validation Test (MRVT)

Circuit Validation Test (CVT)

The Circuit Validation Test (CVT) is designed to ensure that two exchanges have consistent translation data for the circuits connecting the two exchanges.

Figure 8.4 shows an example of how a circuit validation test (CVT) may be performed in an SS7 system:

At the initiating end, an individual circuit to be tested is selected. Tests are then performed to ensure that data exists to:

Derive the physical circuit information for the circuit; and,

Derive the Circuit Identification Code (CIC) from the physical circuit information. If the near-end tests are successful, a CVT message is sent to the remote exchange.

The remote exchange receiving the CVT request message performs the following checks:

Verifies that the CIC indicated in the CVT message is assigned;

Ensures that translation data can derive a physical circuit from the CIC and routing label in the CVT message; and,

Ensures that for the physical circuit, the CIC exists and can be derived from the physical circuit information.

If the far end tests are successful, a CVT message containing the derived CIC is returned to the initiating exchange.

At the initiating exchange, a comparison of the sent and received CICs is mead. IF they match, the test is considered successful.

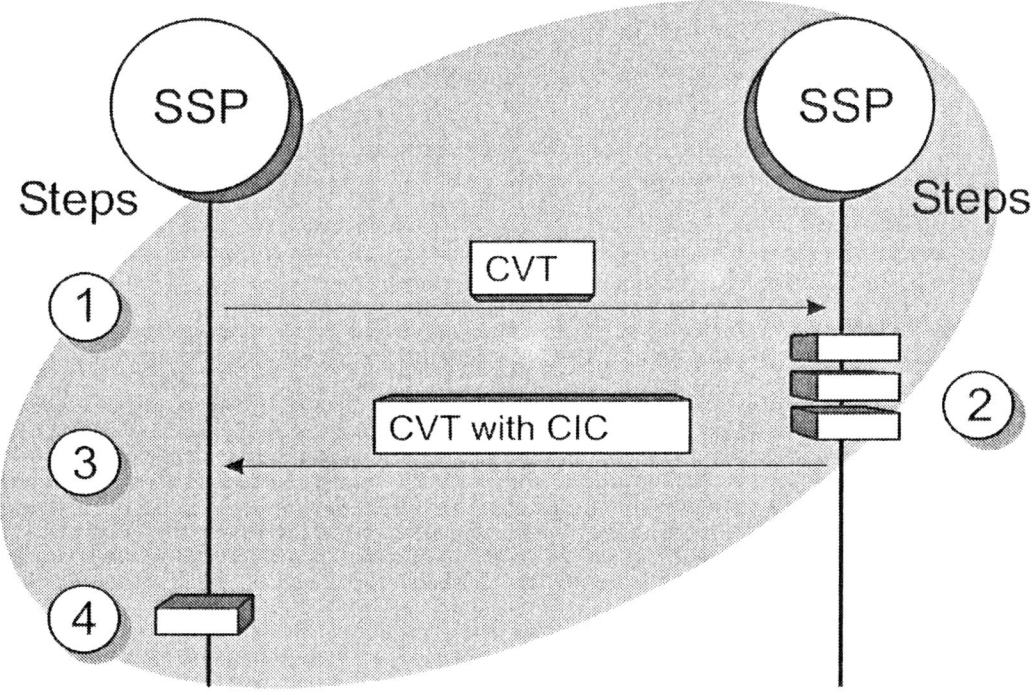

Figure 8.4, SS7 Circuit Validation Test (CVT)

Chapter 9

SS7 Enhanced Services

The ISDN-UP and TCAP sections of this book briefly covered some of the features and enhanced services made possible with SS7 signaling. In this section we look at some of these applications from a broader, end-user perspective. From this perspective, we stand to gain a better understanding of the benefits that SS7 signaling provides for both individuals and organizations.

Until the widespread adoption of SS7, the advanced features the telephone companies developed were limited for use out of one central office; these advanced features are commonly named Centrex. Many private businesses obtained their advanced features and custom applications by purchasing a mini telephone switch for their building: most commonly called a private branch exchange (PBX) or private automated branch exchange (PABX), and lastly stand-alone computer telephony (CT) systems linked buildings together, but not farther than a few blocks apart.

With the introduction of SS7 systems and the Intelligent Network (IN), many new broader applications are possible. The significant advantage of SS7-based applications is realized when the services are offered city wide, county wide, or even nationally. The standardized services now enhanced by SS7 signaling are offered under the common name Custom Local Area Signaling Services (CLASS).

Beyond the local area, SS7 also provides signaling connectivity for enhanced services spanning the nation. Nationwide SS7 provides service intelligence for applications such as Network-based Voice Messaging and Network-based Automatic Call Distribution. Network-based enhanced services all require specific centralized databases containing the unique information for processing the enhanced application.

Figure 9.1 shows the centralized Enhanced Services database residing within the SS7 network that is connected to the local (end office) Service Switching Points (SSPs). Many network-based services utilize Intelligent Peripherals to handle such tasks as calling card digit collection or real-time billing information processing (not storage of billing information). Network-based features also terminate high volumes of call treatments such as voice messaging, status reports, stock reports or weather reports. Adjunct processing devices are used to handle such treatments, although the control of such high-volume applications is handled at an Intelligent Peripheral or an Enhanced Services database. Nationwide 800/toll free service is an example of a network-based enhanced service utilizing a centralized database.

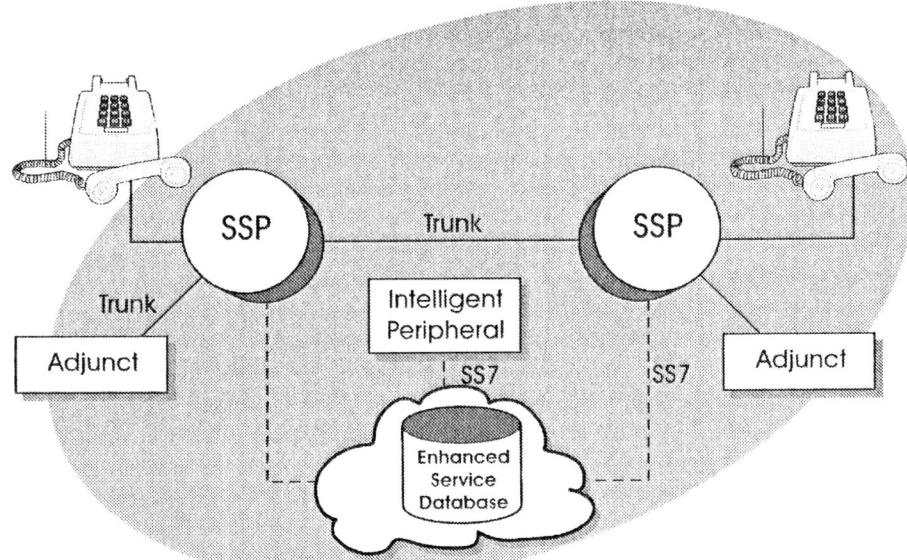

Figure 9.1. SS7 Network Based Enhanced Services Architecture

Network-based Enhanced Voice Messaging

Network-based Enhanced Voice Messaging is a wide area voice messaging system that uses the intelligent routing and control features of the telephone network to route calls to and from voice mail systems. An example of network-based enhanced voice messaging is the automatic routing of calls to a voice mail system that is used by a high tech sales manager who travels between regional offices. Because the sales manager travels constantly, a nationwide messaging system intercepts the call. An Intelligent Peripheral manages the complexities of the call, while the Adjunct device holds onto the call.

- Pressing 1 allows you to leave a message, while
- Pressing 2 activates the call back feature, and
- Pressing 3 routes your call to the sales associate's assistant.

Pressing 1, terminates the call in the Adjunct messaging system nearest the incoming switch, with call handling information (such as Calling line ID) presented by the SS7 network. The sales associate may then automatically call you back upon listening to the message. Your pressing 2, would activate the call back feature, allowing your phone to alert you with a special ring when the sales associate is assumed to be available at one of his office locations. Pressing 3, requests the Intelligent Peripheral to determine which office location the sales associate is working out of, and routes your call to the nearest assistant.

Network Automatic Call Distribution (NACD)

NACD is a is a call processing system that routes (distributes) incoming telephone calls to specific telephone sets or stations calls based on the characteristics of the call or network settings. These characteristics can include routing on network congestion, time of day routing, and other criteria.

NACD is a service that may be used by organizations such as airlines that have customer service centers located across the country, in different time zones. NACD can improve customer service by treating the nationwide pool of agents as if they were all in one location (dialing one phone number). If one NACD location is busy or out-of-service, incoming calls are automatically distributed to pre-determined alternate locations. The airline company as an example, can handle busier hour calls by distributing the incoming call load across several time zones that are not in a busy hour timeframe. The airline gains sales, as more customers get through to more available agents across the nation, not to mention the flexibility the airline has in selecting customer service locations.

While an SS7 message is routed to the Enhanced Services database for instructions, call routing is suspended (for only a matter of seconds) at the incoming SSP. The identified service indicates a complex nation-wide load distribution application is required. The signaling request is then immediately handed off to the Intelligent Peripheral for instructions. Either the call will be immediately routed to the appropriate agent, or in the case of "all agents busy", the call will be routed to the closest Adjunct processor where the caller will be provided with optional automated attendant choices. The call will be completed when the sooner of the Intelligent Peripheral takes back the call for routing or the caller obtained sufficient information from the automated attendant.

Network-based Virtual Private Networking (VPN)

Private Networking provides a broad network user the convenience and simplicity of a private dialing plan (like dialing only 3 numbers for an "in house" call) although the calls are actually routed all over the country. Network-based VPNs share the resources of a larger SS7 interconnected network with a centralized Enhanced Services database. VPNs offer companies the advantage of identical feature operation in different areas of the country while reducing the requirements of owning and maintaining a large private network.

VPN users leverage the intelligence of the network to manage their private dialing plans and long distance agreements. Changes to their dialing plan or new agreements with long distance carriers (as an example) are done at the centralized Enhanced Services database or Intelligent Peripheral.

A Management System (not shown) is required to program the Enhanced Services database or the Intelligent Peripheral. While this diagram shows only a couple switches (SSPs), the connections required for a "private" call setup may actually pass through many switches and/or long distance carriers depending on the dynamically assigned call routing methodology.

Custom Local Area Signaling Services (CLASS)

Custom Local Area Signaling Services (CLASS^{tm}) provides residential subscribers with many of the enhanced features previously available only with PBXs or Centrex service. The flexibility in distance offered by these signaling-based services can provide substantial benefits to busy users. Information contained in the SS7 signaling messages allows for these enhanced services to work beyond the previous limits of within a single rate center or central office boundary.

Figure 9.2 depicts the SS7 network architecture interconnecting multiple local Service Switching Points (SSPs) with a centralized CLASS database. The CLASS database contains personal subscriber information providing necessary routing instructions according to the enhanced feature.

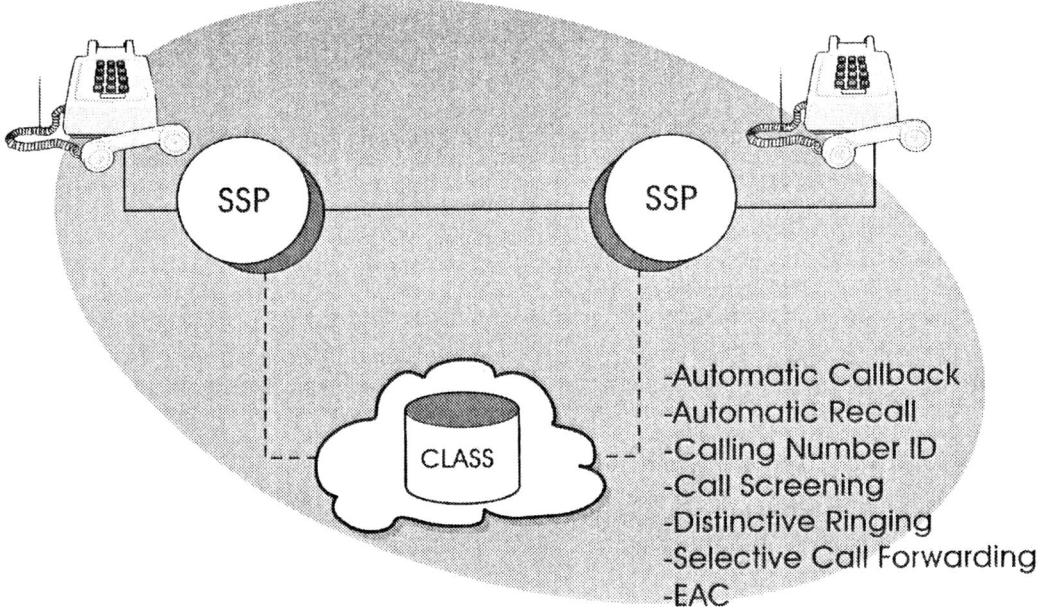

Figure 9.2. Custom Local Area Signaling Services (CLASS) Architecture

The following features describe the call handling parameters and unique feature treatments as managed by the CLASS feature database. Users who subscribe to CLASS features are identified first at the SSP, whereby all calls are then acknowledged by the CLASS feature software.

Automatic Callback

The automatic callback feature allows a caller to request to be automatically alerted to the fact that, the line that was busy is now available. The automatic callback feature involves SS7 signaling between the local switch (caller's switch) and a distant switch (called party's switch.) The automatic callback feature reserves a connection request on the distant (called party's) switch to initiate a connection back to the originating caller when the connection becomes available.

To activate automatic callback service after reaching a number that is busy, the user dials the automatic callback feature code (as per instructions by the service provider) and hangs up the telephone. The originating local switch signals via the SS7 network to the remote switch, an automatic callback request. This reserves (blocks) the called number from receiving additional calls until the automatic callback service has made its first attempt to "call back" the originating caller. When the called number becomes available, the remote switch triggers an SS7 message to the originating local switch and this rings the original caller's phone (with a distinctive ring pattern). The called party's telephone does not ring unless the originator of call back feature responds to the call back alert.

Automatic Recall

Automatic redial, when activated, redials the last incoming call, without the user needing to know the number. Automatic recall is useful if you've ever made a mad dash to the phone, only to find the caller has hung up. Automatic redial to reconnect to the caller without knowing the calling telephone number.

The automatic recall operation starts with the ringing of a call. When the phone rings, an event record is created in that local switch. This event records the connection through the switch containing the calling party number. The calling party's phone number is contained in the Initial Address Message (IAM) that was used to set up the call originally. Then the user dials an automatic recall feature code to initiate the automatic recall service. The local switch reviews the call detail information stored within the event record to determine if the calling party's phone number is available or restricted. If the number is available, then the call is automatically returned to the calling party. This process can occur without the called party knowing the actual phone number of the caller.

Calling Number Identification

Calling number identification displays the calling number during the alerting of an incoming call. Calling number display shows the telephone number (internal or external) between the first and second ring, prior to answering the telephone (many telephones then store the number for later use). The availability of the calling number may also be used to provide for distinctive ringing, selective call forwarding, or selective call rejection.

Figure 9.3 shows the calling number identification operation. Calling number identification operation starts with the reception of a call. When the call is received, the initial address message (IAM) contains the calling party number of the incoming call. The IAM may contain additional information such as the text name of the calling party. This example shows that the local switching system extracts this information and combines this information with the ring signal (using different frequencies and amplitudes) and sends it to the customer during the alerting (ringing) process. If the customer has the appropriate display equipment, the calling number information is display as the telephone rings.

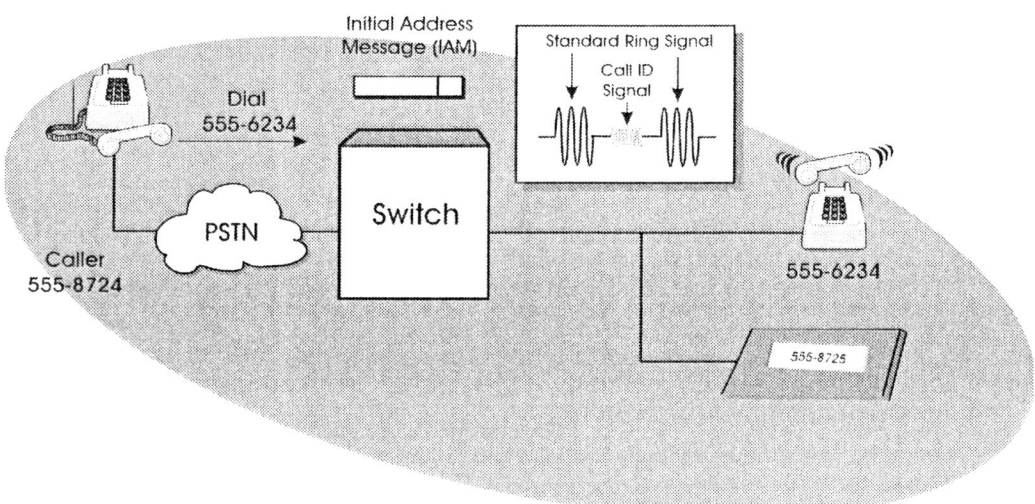

Figure 9.3. Calling Line Identification (CLID) Operation

Calling Number Display Blocking

Calling number display blocking is a feature that gives calling customers the option to change the caller ID presentation status of their number on a continuous or per-call basis. When the calling party's phone number is selected as private, the caller ID display will not be available to the called party. To activate calling number display blocking, the user enters a feature code that sets the presentation restricted field in the SS7 initial address message (IAM) that is sent to the destination (called party's) switch during the call. The IAM may contain the calling party's telephone number, however, this number will not be displayed to the call recipient. The calling party's telephone number will be stored for features such as call trace or automatic call back. When the receiving switch decodes this SS7 message, it alerts the user that the calling number is restricted, private or not available.

Customer Originated Trace (COT)

Call trace allows a subscriber to initiate a call trace request message that allows the phone number of a caller to be stored for later investigation. The activation of the call trace service alerts the telephone service operator to "tag" the originator's number for authorities to investigate the originator of the unwanted or unauthorized call. The call trace activation codes are found in the front of the phone book, or supplied to you in a feature description. If the call trace of the last call was completed successfully, an announcement should be heard. The service operator will then release the call trace information to law enforcement agencies. In some cases a signed authorization form may be required.

Figure 9.4 shows that the call trace operation starts with the reception of an unwanted call. When the call is received, an event record is created in the switch that records the calling party number associated with the incoming call. This example shows that the customer dials a "call trace feature code" to inform the telephone company to trace the last call. The local switch (recipients phone carrier) reviews the call detail information stored in the record to determine if a calling number was provided. If the number was provided and is available, the information will be stored in a special call trace location on the carriers system. The customer or a law enforcement officer then calls the carrier and requests that the number be released for further action. This example shows that the customer completes a call trace release form that authorizes the transfer of the number to authorities.

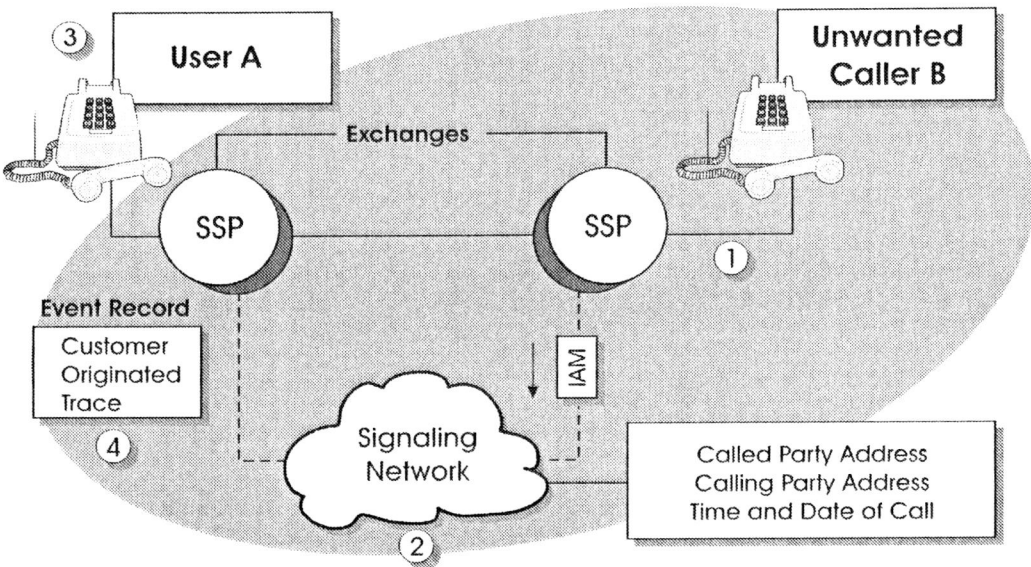

Figure 9.4, Call Trace

Call Waiting

Call waiting is a telephone call processing feature that notifies a telephone user that another incoming call is waiting to be answered while they are currently on the phone. This is typically provided by a brief tone that is not heard by the other callers. An additional feature (which requires accommodating telephone equipment) is "Call Waiting ID" which allows for the telephone number of the calling party that is signaling in, to be displayed on the receiving user's telephone equipment.

After the subscriber is notified of an incoming call (with a short tone), they can either answer or ignore the incoming call. To answer the incoming call the subscriber "flashes" (pressing the on-hook switch for one second) during the current call, and answers the incoming call. While tending to the new call, the caller existing from the original conversation is put in a "call waiting" status. The subscriber may then "flash" (toggle) back and forth between the two calls in progress.

Figure 9.5 shows how SS7 signaling may be used to indicate call waiting service on an analog telephone line. In this example, a call is in process with caller 1. During the call, a second caller (caller 2) dials the telephone number of the user that has call waiting. This results in an initial address message (IAM) being received by the local switch that has CLASS services. The CLASS system discovers that the line or extension is busy on another call. The system also determines that this user has the call waiting service processing feature available so it sends a call waiting message tone to the user (only heard by the user). If the user desires to answer the call, the user sends a flash message (a momentary open on the line) that indicates to the telephone system to place the current call in progress on hold and switch to the other incoming call (caller 2). Each time a flash message is sent, the line alternates between each incoming caller.

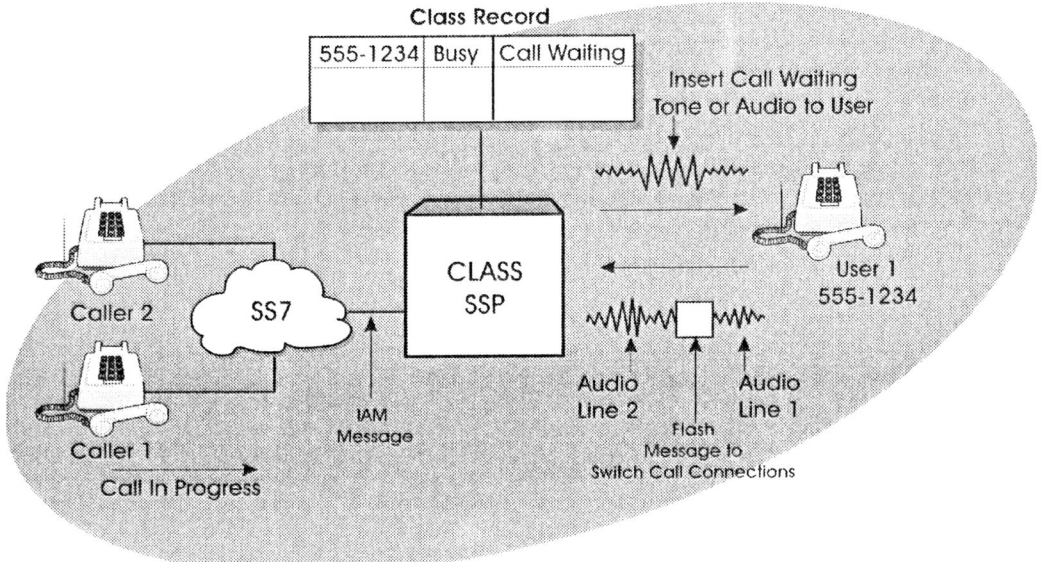

Figure 9.5. SS7 Call Waiting Signaling

Distinctive Ringing

Distinctive ringing is a feature that alerts the subscriber with a special alert (different ringing) sound. The special ring(s) allows for one phone line to be used for multiple purposes. For example business and personal, or urgent verses normal, or even a special ring designated for the fax machine (the fax machine must have distinctive ring capabilities). In any case the calling party (originator) does not know about the distinctive designation.

The distinctive ring is any combination of short and long rings depending on how the feature has been set up by the telephone company. For example: short short long, or short long short, or short short short.

Designation for the distinctive ring can be made two ways:

> As defined by who is calling (calling party's telephone number), and

> Defined by which phone number is called.

Again, in any case the person calling (originator) does not know about the distinctive ring designation.

Designation 1, requires the subscriber to program into the phone system (see the telephone system specific user guide for instruction) a list of "special" phone numbers. When the SS7 network delivers a call with the calling party's phone number in the Initial Address Message (IAM) that matches a phone number on the "special" list, the phone rings a predetermined way indicating a call is arriving from one of the special numbers on the list.

Designation 2, requires the subscriber to have two or more phone numbers assigned to only one physical telephone line. In this case, the phone system identifies the phone number being called as an additional phone number assigned to an original line. SS7 calling party identification is not required for this feature.

Figure 9.6 shows the operation of a telephone system that has distinctive ringing feature. It shows a single telephone that is assigned two different telephone numbers even though the telephone can operate on only one telephone line (one switch port). This example shows an incoming call is received for the registered number 555-6234, it is re-directed (forwarded) to the original destination phone number 555-1234 along with information that allows the system to uniquely identify the call with a distinctive ring (for example short, short, short). Although not shown a third telephone number may be assigned to the same line and programmed to ring in a different way altogether (for example long, long, long). This can be repeated with still yet more phone numbers assigned to the original line. When calls are received to the original 555-1234, the ring is the standard long, long cadence. Thus allowing the receiver of the call to know in advance which telephone number was dialed by the calling party.

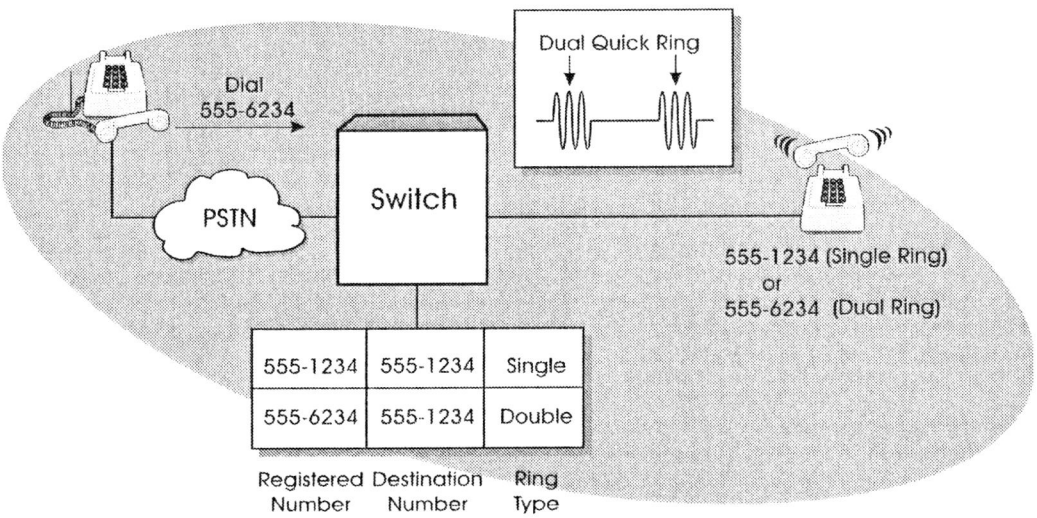

Figure 9.6. Distinctive Ringing

Selective Call Forwarding

Selective call forwarding is the processes that transfer incoming calls (identified by telephone number) to a different telephone number based on one or more criteria. Selective call forwarding is a service that requires the calling party information transmitted in the SS7 Initial Address Message (IAM). Selective call forwarding can be used to redirect important phone calls to your wireless phone (for example) while all other calls continue to dialed number destination. An example scenario might be to add a person's boss to the selective call forwarding list while out to dinner, so all calls (including solicitation calls) ring the home telephone number except the one (or more) important call that is (re-routed) forwarded to the users wireless phone.

Figure 9.7 shows how selective call forwarding can be used to deliver calls to alternate number based on a specific criteria type. This diagram shows a selective call forwarding service that routes fax calls to different telephone number or extension after it detects the call is coming from a number on the selective call forwarding list (a fax in this case). After the system detects that the incoming call is "on the list" the local telephone switch transfers the call to the destination number assigned by the selective call forwarding feature.

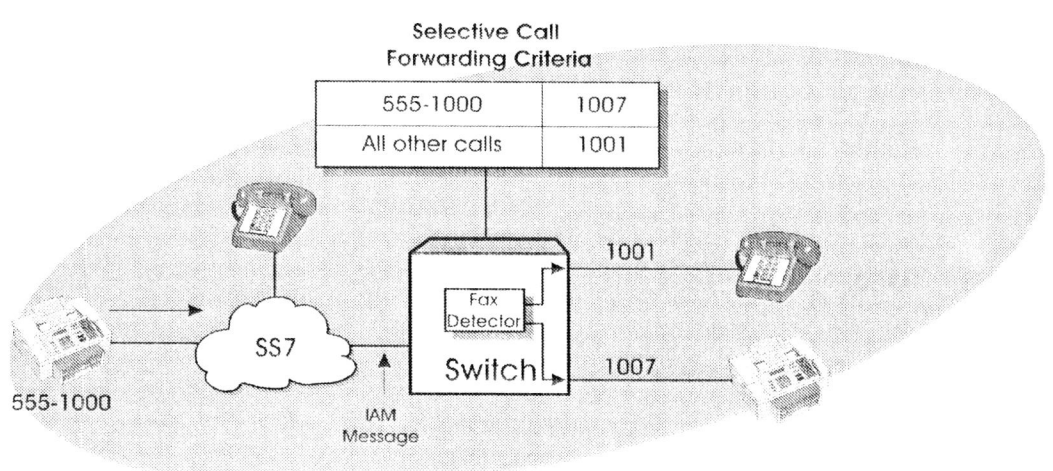

Figure 9.7. SS7 and Selective Call Forwarding

Selective Call Rejection

Selective Call Rejection is used to block calls from known or unknown undesired callers. Rejected calls are provided with a pre-recorded announcement that states the number is not accepting calls. The feature description manual provides instructions on how to add the last unwanted call onto the list (without knowing the number) or how to add a specific known phone number to the list.

Selective call rejection is a feature much like Selective Call Forwarding, except that it restricts the delivery of certain calls to their dialed telephone number destination. This feature also uses a list that is programmed by the subscriber.

Figure 9.8 shows how selective call rejection can be used to restrict the delivery of calls from a specific list of callers. This diagram shows that a call from a harassing caller is added to the rejection list. The caller will then be routed to an automated message unit that plays a pre-recorded announcement of not accepting calls.

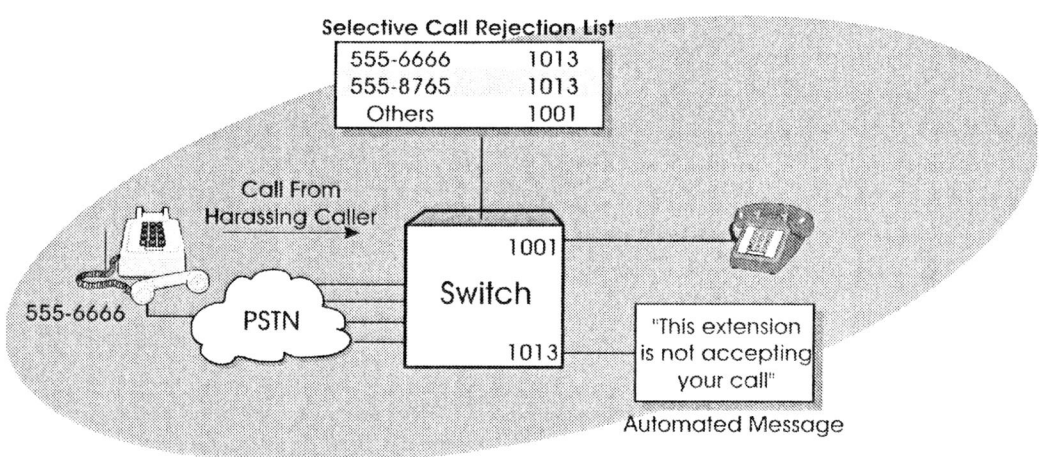

Figure 9.8. Selective Call Rejection

Enhanced 800 Services

Enhanced 800 service is a call service option that allows for the routing of 800 (or other toll free or numbers such as 888, 877...) to be routed to different location based on other criteria such as day of week, time of day, or caller location. SS7 messaging is required before the call can be completed. Messaging to the "800" database requires the calling party's phone number should the routing be based on the callers location. Because the toll free service is a national service, originating location is especially desirable. For roadside assistance for example, calling a national number and reaching someone across the country provides a disservice to the caller.

With the time-of-day and day-of-week routing options, organizations can set up full-time customer service centers backed up by part-time centers which only operate during peak business hours. All calls after hours, or on weekends, would automatically be routed to the main business office for special instructions.

Enhanced 800 service may also be restrictive in nature. A call based on location of caller provides the call to be routed based on other parameters such as the area code or exchange code (the first 3 digits after the area code is called the exchange or office code). An organization could then restrict an 800 number to be valid only in certain cities or regions within a state. When all these parameters are combined, Enhanced 800 provides a powerful and flexible way to meet certain telecommunications requirements.

Figure 9.9 shows an example how an Enhanced 800 number translation service operation allows a company to route calls to different service centers depending on the time of day and location of the caller. This diagram shows that a company has 3 offices; Boston, San Francisco, and Dallas and the company have a single toll free line 800-227-9681. Callers from east coast exchanges are routed to the Boston office from 9:00am – 5:00pm EST. Callers from west coast exchanges are routed to the San Francisco office from 9:00am – 5:00pm PST. All other calls are routed to the Dallas office. This allows the corporate office to handle calls after the regional offices are closed.

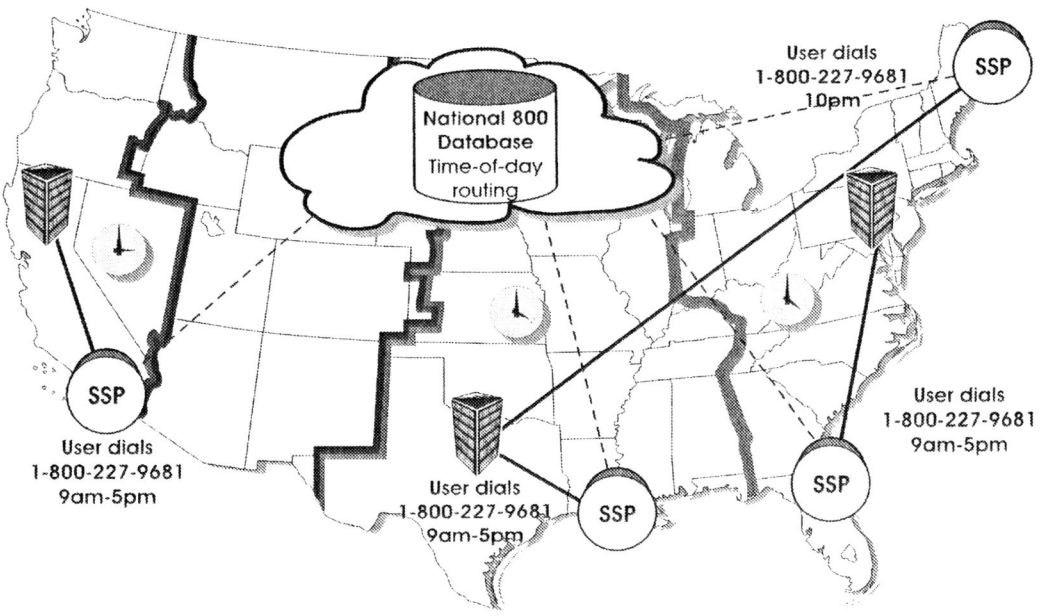

Figure 9.9, Enhanced 800 Service

Calling Card Services

Calling card services allow the customer to use identifying numbers or codes from calling cards to uniquely identify and individual or account. These services enable the individual to be separately charged for the various usages on the account. Calling cards may be post pay (charged after the calls are made) or prepaid (paid for a certain amount of time by the customer in advance).

The Signaling System 7 network is uniquely valuable for these services to minimize the time it takes to transfer various legs of the call and transfer valuable (billing) information with the call. A prepaid calling card is issued or activated by a telecommunications service provider that provides coded identification information permitting the card holder to initiate a call.

Prepaid calling cards use numbers stored on a magnetic stripe or the identification code numbers may be dialed by the user that uniquely identifies the card and authorizes access to the telephone system. Prepaid calling cards require real-time billing to end a call if the time limit has been exceeded.

Figure 9.10 shows how the SS7 system can be used to allow for calling card operation. This example shows the use of a prepaid calling card. The call is routed to an Adjunct processor where the digits are collected either prompting the user to enter information or information from the magnetic stripe is sent by the electronic phone. The Adjunct then sends the account information (dialed digits and account number) to the real time rating system located in the Intelligent Peripheral. The real-time rating system identifies the correct rate table (e.g. peak time or off peak time) and inquires the Enhanced Services database for the account balance associated with the calling card. Using the rate information and balance available, the real time rating system determines the maximum available time for the call duration. This information is sent back to the Adjunct to complete (connects) the call. During the call progress, the Adjunct maintains a timer so the caller cannot exceed the maximum amount of time. After the call is complete (either party hangs up), the Adjunct sends an SS7 message to the real time rating system containing the actual amount of time used. The real time rating system uses the time and rate information to calculate the actual charge for the call. The system then updates the Enhanced Services database with the subscribers account balance (decreases by the charge for the call).

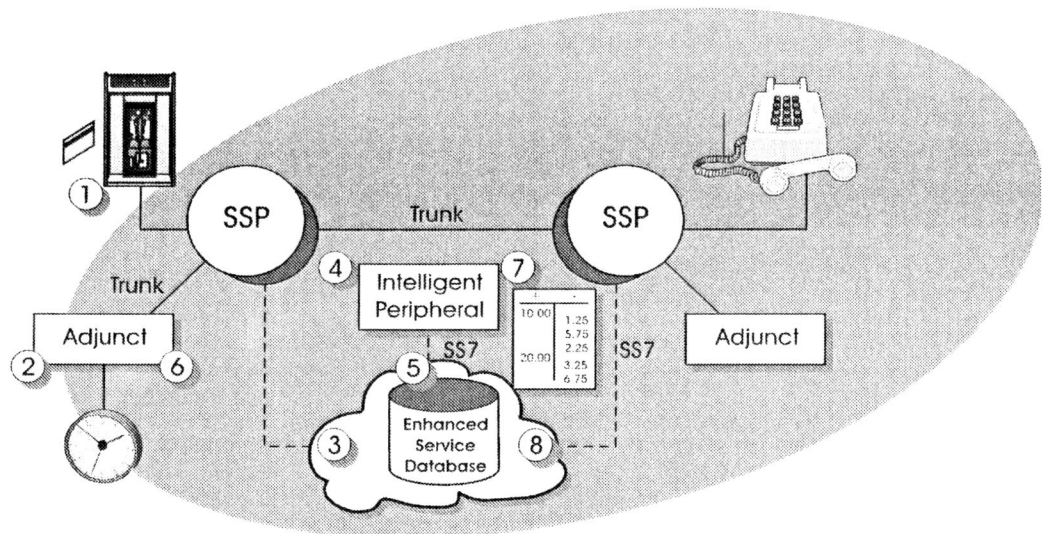

Figure 9.10. Prepaid Calling Card Network Architecture

SS7 Network Databases

There are many different types of network databases accessible by the SS7 Network. Most of the key revenue generating services (such as 800 /toll free calling) utilize network databases for information translation.

Service providers use SS7 Network databases for subscriber information, billing information, equipment information (wireless handsets), or special services translation such as local number portability.

Below is a list of common SS7 Network Databases:

- Service Control Point (SCP)
- Line Information Database (LIDB)
- Emergency 911 database (E911)
- Enhanced 800 database (800)
- Local Service Management Systems (LSMS)
- Authentication Center (AUC)
- Home Location Register (HLR) wireless
- Visitor Location Register (VLR) wireless
- Equipment Identity Register (EIR)
- Billing Center (BC)
- Calling Card database (CCDB)
- Intelligent Peripheral (IP) enhanced services processor

Chapter 10

Mobile Application Part (MAP)

The Mobile Application Part (MAP) specifies the application protocols between Mobile Switching Centers (MSCs) and related SS7 network equipment assemblies and databases that are used in wireless networks.

Figure 10.1 shows how the MAP functional part is located within the SS7 system. This diagram shows that the MAP functional is part of the TCAP function and that the MAP functional part in the SS7 system is equivalent to the OSI network model application layer 7.

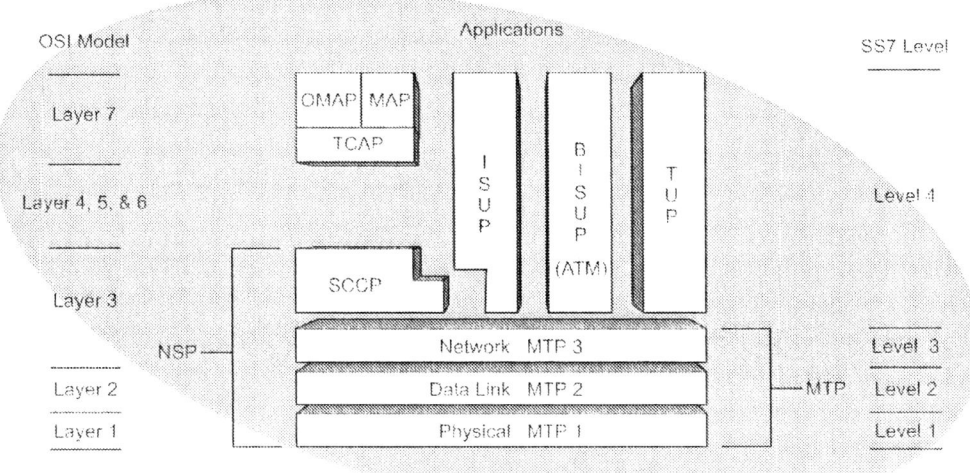

Figure 10.1. MAP in the SS7 Protocol Stack

Wireless Networks

Wireless networks inter-connect wireless telephones with nearby radio towers that route calls through switching systems to other wireless telephones or to other telephones or data networks. Wireless networks consist of cell site radio towers (called "Base Stations"), communication links, switching centers and network databases. Wireless networks usually interconnect to public telephone and data networks such as the Internet.

The main switching system in the mobile telephone service, wireless network is the mobile switching center (MSC). The MSC coordinates the overall allocation and routing of calls throughout the wireless system. Inter-system connections can link different wireless network systems together to allow wireless telephones to move from cell site to cell site and system to system. The mobile system defines inter-system connections in detail to allow universal and uniform service availability for mobile wireless devices.

Figure 10.2 illustrates the fundamental parts and interconnections in a mobile communication network and how some of them use the SS7 system. The mobile station (MS) is a wireless telephone that communicates with nearby radio towers. The radio towers are composed of an antenna system and a base station. The base station is found near the base of an antennae structure and is composed of radio base station transceiver sub-systems called Base Transceiver Station (BTS) and Base Station Controllers (BSCs). The BTS convert radio signals (see Um in figure) from mobile station and converts (transcodes) the information into a form suitable for transfer to the public network. The BSC coordinates the overall operation of the BTS units. The BSC units are coordinated by mobile switching centers (MSCs) and subscriber databases.

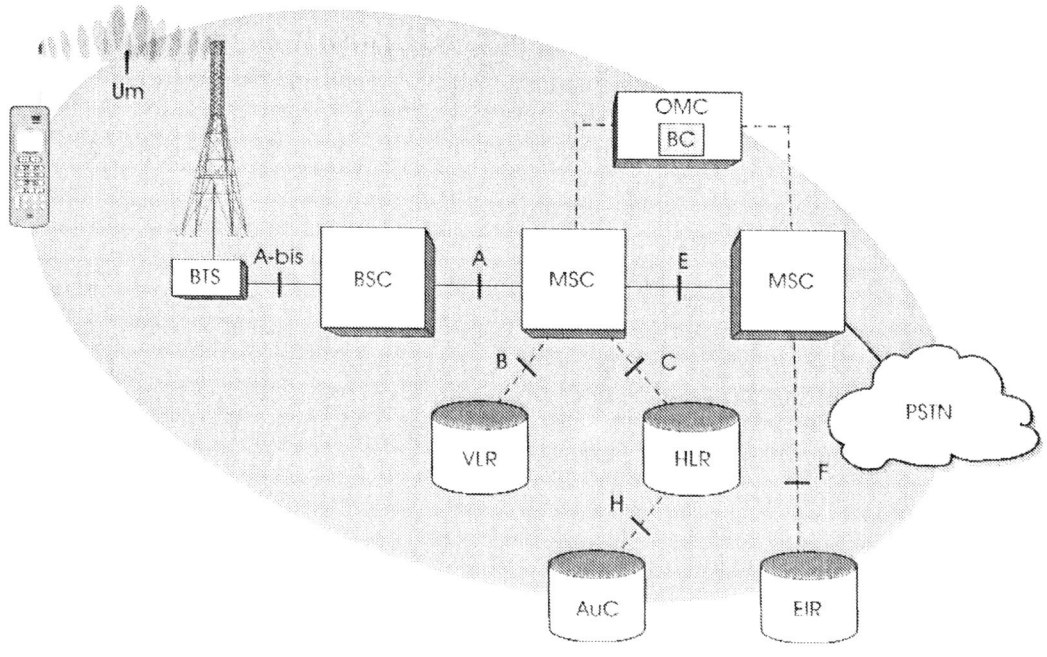

Figure 10.2. Mobile Communication Network and SS7

The core network interconnects the mobile communication system mobile switching centers (circuit switches) and general packet radio services switches (packet routers) to other networks such as the PSTN and the Internet. The MSC coordinates a majority of the voice service processes. Regardless of the type of radio transmission, the MSC routes calls to and from cell sites and the PSTN.

Some mobile communication systems include a packet switching network as an integrated part of its network. For example, the general packet radio service (GPRS) includes packet switches that are connected directly to radio resources. These packet switches can be divided into a serving node that coordinates an active wireless data device and a gateway node that provides interconnection with a data network. The wireless data device may actually be a wireless telephone that has data capability.

Mobile communication systems have interfaces between network parts. Some of these interfaces are "open interfaces." Open interfaces are defined in detail that allows network equipment that is produced by different manufacturers can inter-operate with each other. Open interfaces are successfully used around the world, and as a result, there are more manufacturers of network equipment parts and lower equipment costs due to market competition.

Base Transceiver Station (BTS)

Base stations may be stand alone transmission systems are part of a cell site and is composed of an antenna system (typically a radio tower), building, and base station radio equipment. Base station radio equipment consists of Radio Frequency (RF) equipment such as transceivers and antennas, controllers, communication interfaces, and power supplies. Base Transceiver Station (BTS) have many of the same functional elements as a wireless telephone. However, base station radios are much more powerful and are solely coordinated by the MSC/BSC and have many additional functions than a mobile telephone.

Also in Figure 10.2, communication links can be seen (A and A-bis) which carry both data and voice information between the MSC and Base Transceiver Station (BTS). Options for the physical connections include wire, microwave, or fiber optic links. Alternate communication links are sometimes provided to prevent a single communication link failure from disabling communication. Regardless of the physical type of communication link, the channel format is usually the same. Communication links are typically digital time-multiplexed to increase the efficiency of the communication line. The standard format for time-multiplexing communication channels between cell sites in North America is the 24 channel T1 line, or multiple T1 channels. The standard format outside of North America is the 32 channel (30 useable channels) E1 line.

Base Station Controller (BSC)

BSCs coordinate the overall operation of each the Base Transceiver Station (BTS) assigned to it. The BSC manages the connections of mobile telephones that are operating within the reach of its own radio subsystems. This means tracking phones and wireless data devices for necessary handoff activity.

The BSC performs control signaling routing and message processing from commands that are received from the MSC (see A interface in figure 10.2). The BSC also adds additional control commands such as handoff messages. Base station controllers insert control channel signaling messages, set up dedicated channels (for voice and data), and monitor the channels for channel quality levels that may indicate a need for handover. In addition, controllers monitor equipment status and report operational and failure status to the MSC.

Switching Centers

A switching center coordinates all communication channels and processes. There are two types of switches used in mobile communication systems; a mobile switching center (MSC) and a general packet radio services (GPRS) support node.

The switching assembly connects the base station subsystems (BSS) as seen in figure 10.2 and other networks such as the PSTN or the Internet. Early analog switches required a physical connection between switch paths. Today's switches use digital switching assemblies that are high-speed matrix memory storage and retrieval systems. These systems provide connections between incoming and outgoing communication lines.

Mobile Switching Center (MSC)

The Mobile service Switching Center (MSC), formerly called the mobile telephone switching office (MTSO), processes requests for service from wireless telephones and land line callers, and routes calls between the base stations and the PSTN. The MSC receives the dialed digits, creates and interprets call processing tones, and routes the call paths.

A system controller coordinates the MSC's operations. A communications controller adapts voice signals and controls the communication links. The switching assembly connects the links between the base station subsystem (BSS) and the PSTN. In support of disaster preparedness, the systems have power supplies and backup energy sources to power the equipment.

There are two types of MSCs; the serving mobile switching center (MSC) and the gateway mobile switching center (GMSC). This is the logical separation of the MSC for directly controlling the mobile telephone (SMSC) and providing a bridge between other networks such as the PSTN (GMSC).

Serving Mobile Switching Center (SMSC)

The serving mobile switching center (SMSC) is the switch that is connected to the BSC that is providing service directly to the mobile telephone. The SMSC is responsible for coordinating the transfer of calls between different BSCs. When the call is transferred to BSCs that are connected to a different MSC, the role of SMSC will be transferred to the new MSC. In GSM systems, the original switch remains in the call after handoff to a second switch and is therefore known as the anchor MSC.

Gateway Mobile Switching Center (GMSC)

The gateway MSC (GMSC) is the point where the wireless network is connected to the public circuit switched telephone networks (typically the PSTN). All PSTN call connections to a mobile (mobile terminated calls) must enter the network at a GMSC, but calls may leave the network from any switch, not just a GMSC. The GMSC maintains communication with the SMSC as calls are transferred from one system (or different MSCs within a system) to another.

General Packet Radio Service (GPRS) Support Node

GPRS is a separate switching system used for the transfer of packet data. This packet switching network is composed of different types of support nodes that receive and transfer packets towards their destination. The general packet radio service (GPRS) support node performs a similar function as the MSC except it switches packets instead of maintaining a specific call connection path. Packet nodes used in the network are divided into serving support nodes and gateway support nodes.

Serving GPRS support node (SGSN)

The serving GPRS service node maintains packet data communication with the mobile telephone via the radio network. The SGSN will sense, register and maintain information about packet data radios operating in its radio network. As the mobile station (telephone) moves through the system, the SGSN will ensure packets are routed to the base station closest to the mobile station (MS).

Gateway General Packet Radio Service (GGSN) Support Node

The gateway GPRS support node (GGSN) is a packet switch that routes packets between the core network and external data networks such as the Internet. The GGSN is the inter-working function that is responsible for adapting and buffering the information between the various networks.

Network Databases

There are many network databases used in mobile communication networks. Some of the key network databases include a master subscriber database (home location register), temporary active user subscriber database (visitor location register), unauthorized or suspect user database (equipment identity register), billing database, and authorization and validation center (authentication).

Home Location Register (HLR)

The home location register (HLR) is a subscriber database containing each customer's international mobile subscriber identity (IMSI) to uniquely identify each customer. Local wireless carriers require only one HLR, although national carriers may have several HLRs. The number of HLRs used within a network is determined primary by the number of subscribers registered to a network. This is because each primary database (HLR database) will have a limit to the number of records it can store. As seen in figure 10.2, the defined interface between the HLR and the MSC is called the "C" interface.

The HLR holds each wireless customer's entire user profile. This profile includes the selected long distance carrier, calling restrictions, service fee charge rates, and other selected network options. The subscriber can change and store the changes for some feature options in the HLR (such as call forwarding). The MSC system controller uses this information to authorize system access and process individual call billing.

Most HLRs are large fully redundant, high availability or fault tolerant database systems. Subscriber databases are critical, so they are regularly backed up in the rare case the need arises to restore the information should the HLR system fail.

Visitor Location Register (VLR)

The visitor location register (VLR) contains a subset of a subscriber's HLR information for use while a mobile telephone is active on a particular MSC. The VLR holds both visiting and home customer's information. The VLR eliminates the need for the MSC to continually check with the mobile telephone's HLR each time access is attempted. The HLR information is temporarily stored in the VLR memory, and then erased either when the wireless telephone registers with another MSC or in another system or after a specified period of inactivity.

Equipment Identity Register (EIR)

The equipment identity register is a database that contains the identity of telecommunications devices such as wireless telephones and personal digital assistants (PDAs). The status of these devices in the network is set to authorized or not-authorized. The EIR is primarily used to identify wireless telephones that may have been stolen or have questionable usage patterns that may indicate fraudulent use. The EIR has three types of lists; white, black and gray. The white list holds known good IMEIs. The black list holds invalid (barred) IMEIs. The gray list holds IMEIs that may be suspect for fraud or are being tested for validation.

Billing Center (BC)

A separate database, called the billing center, keeps records on telephone usage. The billing center located within the Operations and Maintenance Center (OMC) as seen in Figure 10.2, receives individual call records from each MSC and other related network equipment. The switching records (connection and data transfer records) are converted into call detail records (CDRs) that hold the time, type of service, connection points, and other details about the network usage that is associated with a specific user identification code. The format of these CDRs includes flexible billing record formats for voice and data usage. These billing records are then transferred to a separate computer by electronic data interchange (EDI). The billing system or company can then invoice subscribers and settle bills between different service providers (a clearinghouse company).

Authentication Center (AuC)

The Authentication Center (AuC) stores and processes information that is required to validate the identity ("authenticate") of a wireless device before service is provided. During the authentication procedure, the AuC processes information from the wireless device (e.g. IMSI, secret keys) along with a random number that is also used by the mobile telephone to produce an authentication response. The AuC compares its authentication response results to the authentication response received from the mobile telephone. If the processed information matches, the wireless device passes. The need for this level of authentication is desirable with the proliferation of new wireless Internet and data sensitive applications such as stock trading.

Public Switched Telephone Network (PSTN)

The Public Switched Telephone Network (PSTN) is the landline telephone system that connects a wireless telephone to any telephone in the world. Wireless telephones, landline telephone service, and various other networks such as private automatic branch exchanges (PABX), all have unique interface requirements. The standardization of SS7 has allowed the beginning of feature transparency between wireless and wireline providers. See Chapter 9 for an example of some of these SS7 based features.

Internet

The Internet is a network of networks that can understand a common data communication language. Although each network within the Internet may talk different language between their elements (data nodes and switching points), they can receive and forward packets through their network to their destination. This is accomplished by a standard set of addressing and routing rules (protocols).

The most common protocol is the Transaction Capabilities Protocol (TCP) combined with the Internet Protocol (TCP/IP). TCP involves the tracking and confirmation of packets sent and received through the Internet. IP is only concerned with addressing and routing packets. The Internet is primarily composed of routers that receive and forward packets toward their destination. Routers are smart switches that dynamically learn where to send packets they receive. Routers are initially programmed with routing tables that indicate where to send packets.

Unlike the fixed addressing of SS7, the IP routers are connected to other routers, which automatically inform its neighboring routers of its presence and address. This automatic addressing continues with routers updating their routing tables with the information and broadcasting their new information to other routers. Over short period of time, many routers in the network have updated their routing table information and packets will be more efficiently forwarded to their destinations.

Mobile Application Part (MAP) Interfaces

MAP protocols have been enhanced to provide for mobility management signaling and basic services support. Specific versions of MAP are used between different network elements (e.g. between a MSC and a VLR). MAP protocol supports radio resource management, mobility management, connection management, services support, and short messaging service (SMS).

Figure 10.3 is a functional block diagram of a wireless network and how it uses MAP protocols between the equipments. This diagram shows that the different versions of SS7 MAP are used between network elements in a wireless network. It also shows that MAP is not used in the radio link. Instead, the relative parts of MAP are transformed into commands that can be sent on the radio links.

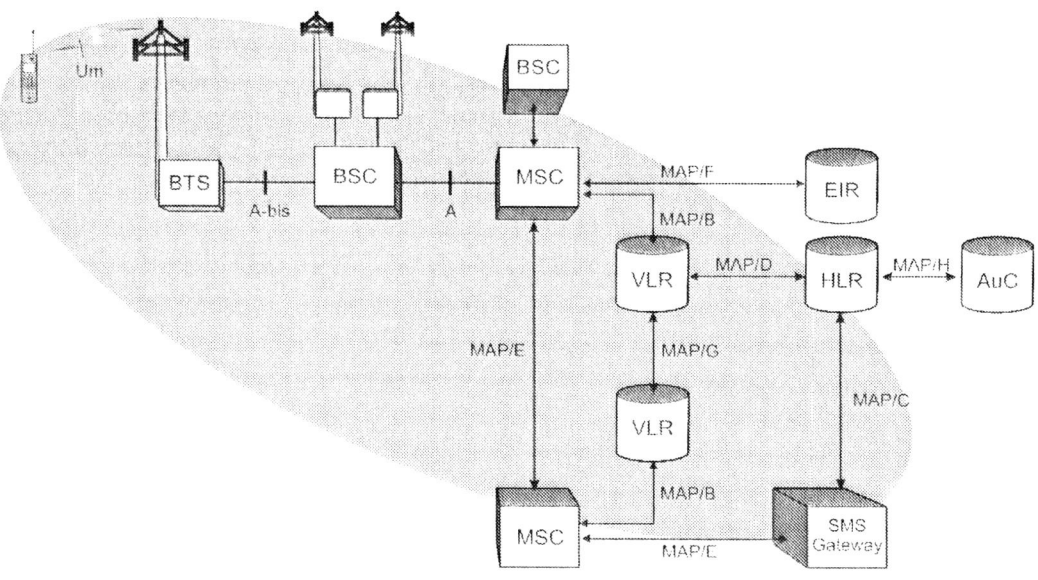

Figure 10.3. SS7 MAP Network

Mobility Management

Like the advanced services found in landline networks, mobile networks also utilize the SS7 network for call control. SS7 can be used to quickly transmit information between MSCs and to query mobile databases for supplementary services.

Figure 10.4 shows how a mobile phone may register with a visited network database that is connected by the SS7 system to the mobile phone's home system. This example shows that the mobile phone will register with the mobile network when it detects a signal from a new system (usually based on the system identification code received from the cellular system broadcast message). This registration is sent from the BTS radio transceiver to the BSC that forwards it onto the MSC (step 2). To discover the authenticity and features assigned to this mobile phone, the MSC sends an SS7 message to the mobile databases that are updated using the Mobile Application Part (MAP) protocol specification. The network database receives the request and replies with validation information that should authorize service for the mobile customer. Because the home MSC is connected to the visited MSC by the SS7 system, this registration allows the home MSC to automatically forward calls to the visited system.

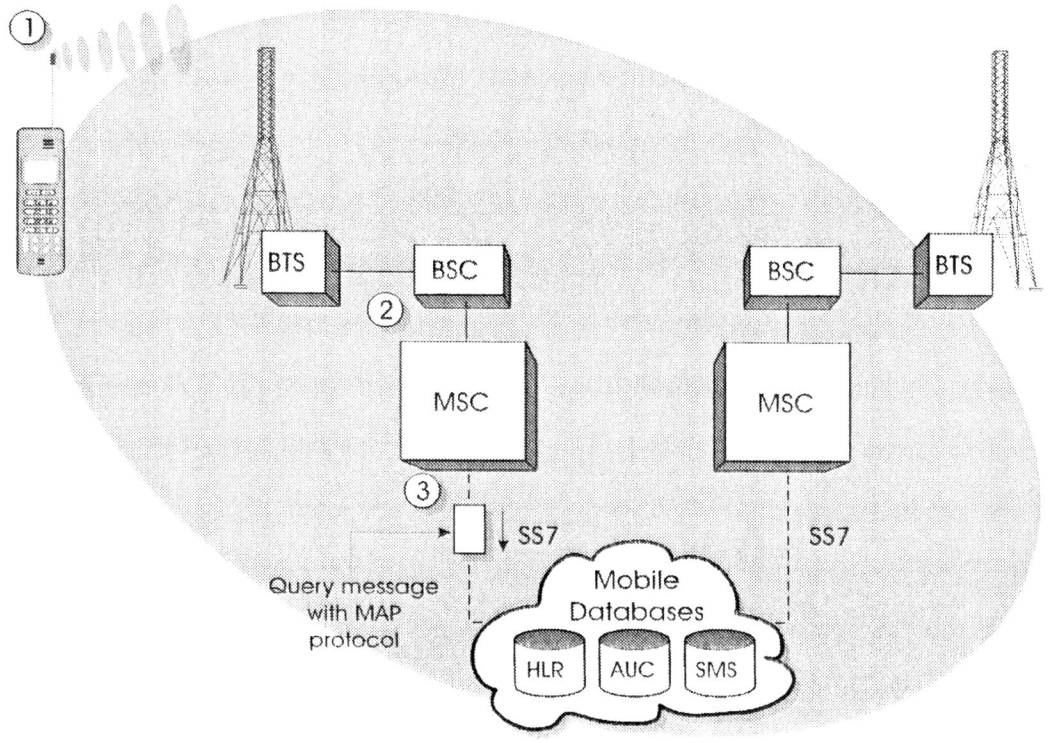

Figure 10.4. Mobile Registration using SS7

SS7 Applications of Mobility

Wireless service providers that support the SS7 signaling Mobile Application Part (MAP) protocol are able to provide (for example) the following advanced mobile services:

> Mobile Station (MS) Position Capabilities such as 911 location service or proximity services such as the nearest hospital to the mobile user. This class of service utilizes the Location Service specification of the MAP protocol that specifies the inquiry information for all geographical position based enhanced services.

Text Messaging supplied by a network-based, Short Message Service (SMS) platform. The SMS platform transfers messages that are ultimately carried utilizing the MAP protocol. The SMS platform may receive input from variety of sources such as:

- an operator assisted service
- a mobile phone
- an Internet Protocol (IP) gateway
- emergency response service
- other one-way, automated text-based systems

Service Session Capabilities such as a two-way interaction Internet session. These service sessions require a sophisticated mobile phone (or device) to support bandwidths much faster than the short message service text messaging.

Figure 10.5 shows how the SS7 network will route specific messages holding Mobile Application Part (MAP) data to and from network databases designed for mobile applications. An example of the mobile application databases are as follows: an Equipment Identity Register (EIR) for verification and security purposes related to the type or ownership of the phone equipment or mobile device, Short Message Service (SMS) store-and-forward database to hold messages for confirmation or broadcast, and a Home Location Register (HLR) which stores subscriber billing and service profile information.

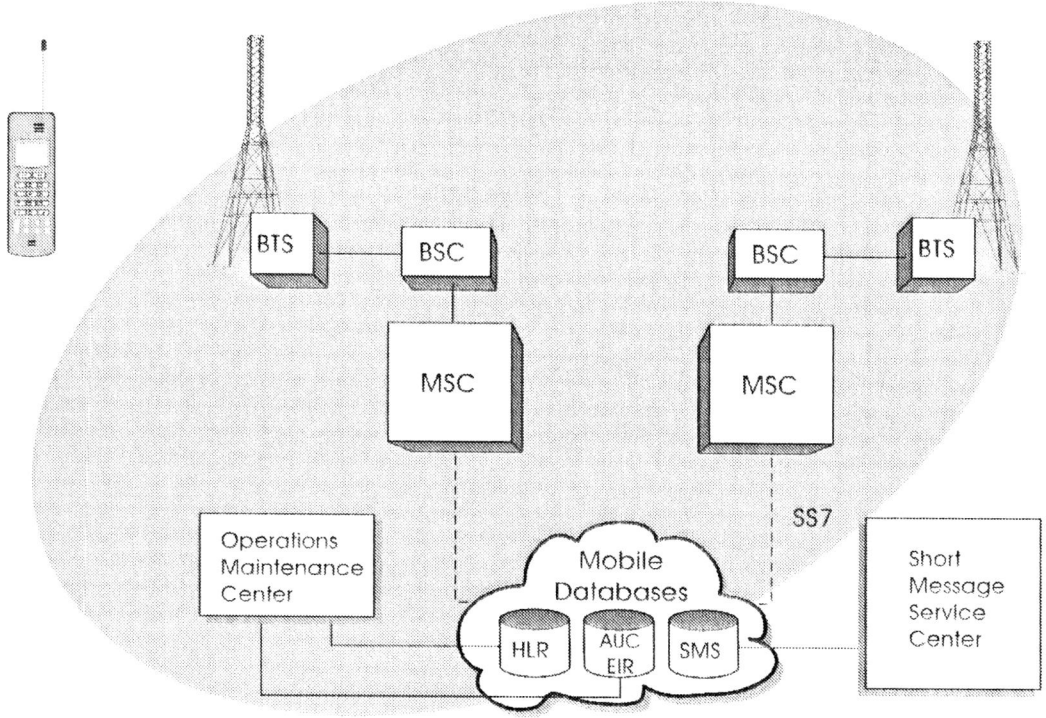

Figure 10.5, Mobile Network Database Access using SS7

Figure 10.6 shows an example of how the call flow for a short message service may be sent from one mobile phone to another mobile phone. This example shows that short message delivery is a store and forward service that can be processed even when the recipient device is unavailable (such as shut off in this example). This example shows that mobile 1 sends a short message request to the MSC. The MSC contact the SMS center and transfer the message. The SMS center validates that is can deliver the message to the destination address and provides a positive acknowledgement back to the MSC that it has received the message and will eventually deliver the message to the recipient. The SMS center then contacts the MSC of the message recipient to deliver the message. Unfortunately, the MSC informs the SMS center that the recipient is not available (NACK response) so the SMS center informs the HLR that it has messages waiting for the recipient. When mobile 2 (the recipient) turns on, it registers with the system. This registration message is received by the HLR. Because the HLR knows it has messages waiting for the recipient, it informs the SMS that the recipient's mobile is now available. The SMS will then send the SMS message to the MSC that delivers it to the recipient. When the recipient has received the message, it will acknowledge it to the MSC and the MSC will confirm its delivery to the SMS center.

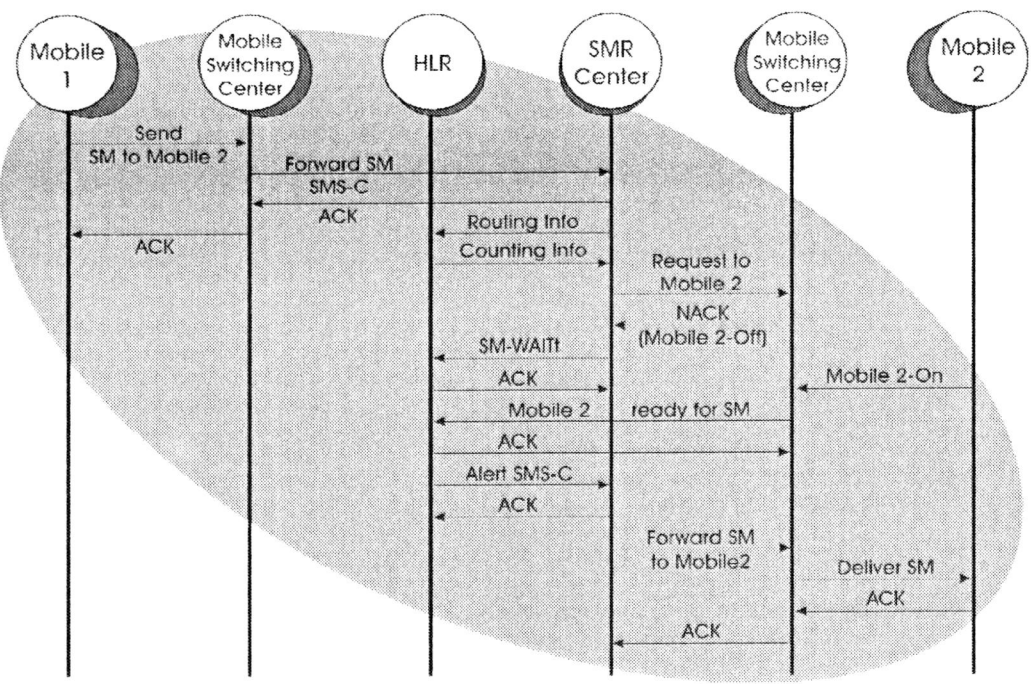

Figure 10.6, Short Message Delivery Signaling using SS7

Basic Service Support

MAP protocols have been developed to provide for basic services support. This includes the setup and connection management of all mobile calls. Examples of basic service support include paging, connections, and authentication of mobile telephones.

Mobile Paging

Mobile paging is the process of locating a mobile telephone and alerting the device of an incoming call or service. Information about the location of the mobile device relative to the nearest radio base station antennae is retrieved periodically such as when the mobile device is powered on. This location information is updated in the Visitor Location Register (VLR) database.

Figure 10.7 shows an example of how an incoming call from the PSTN can alert (page) a mobile device in a visited system through the use of SS7 signaling. This diagram shows that a caller dials the telephone number of the mobile phone. When the incoming call is received by the MSC of the home system, it sends a message to the VLR to determine if the mobile is active in its home system. When the MSC determines that it is mobile phone is registered in a visited system, it sends a message to the visited system informing it of an incoming call. The visited system then searches its VLR to determine where within its network that the mobile phone has last registered. The MSC will then send an alert message to the radio tower(s) where the mobile is expected to be operating and this message is sent as a paging message on the radio channel.

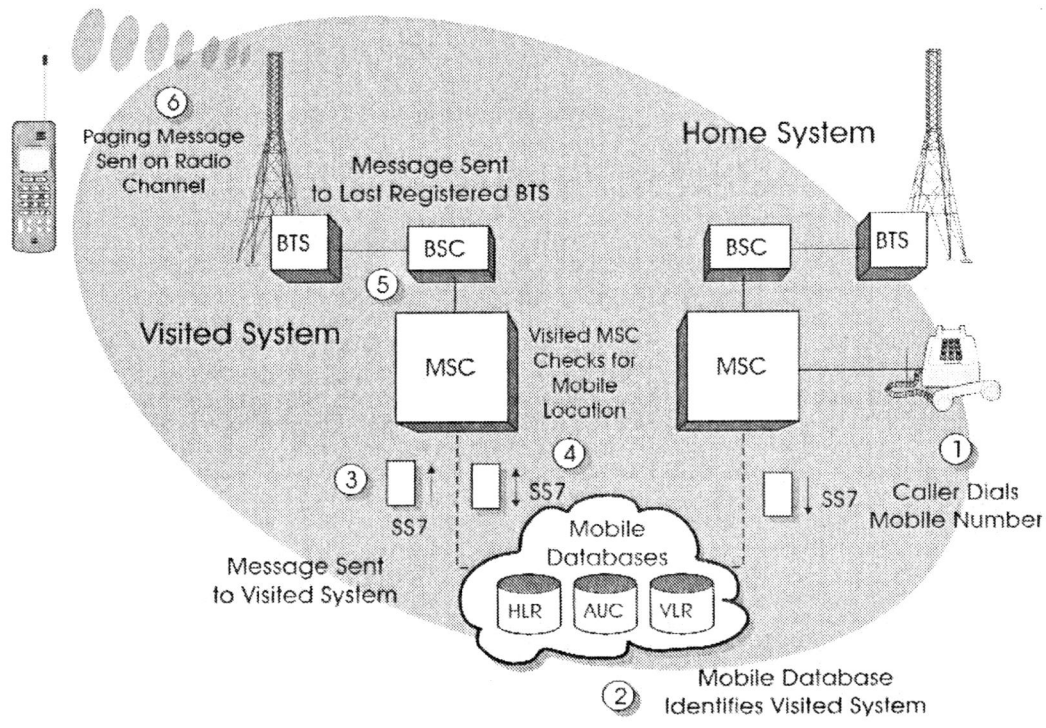

Figure 10.7. SS7 Mobile Paging Signaling

Authentication

Authentication is a process during where information is exchanged between a communications device such as a mobile phone or a personal digital assistant (PDA) and the SS7 based authentication database. This allows the wireless carrier to confirm the true identity of the device requesting service. Original fraud detection systems were poor in identifying fraudulent activity because the detection was done far after the initial access to the network. Utilizing SS7 signaling, a real-time validation of the authenticity of the device can be accomplished. This allows a service provider to deny usage of the device before fraudulent calls can be made (e.g. shortly after the device is turned on).

Figure 10.8 shows the Authentication Center (AuC) being queried through the SS7 network immediately after mobile phone has requested to initiate a telephone call in a visited system. This example shows that the mobile radio has or will receive a random number from the mobile system. It uses this random number along with other information to calculate a response (SRES). This SRES is sent to the mobile system. The visited MSC contact the authentication center (AuC) associated with the mobile phone and sends the random number and other information to the AuC. The AuC uses this information to calculate a response (SRES). If the SRES from the AuC and the mobile phone match, the MSC can complete the call.

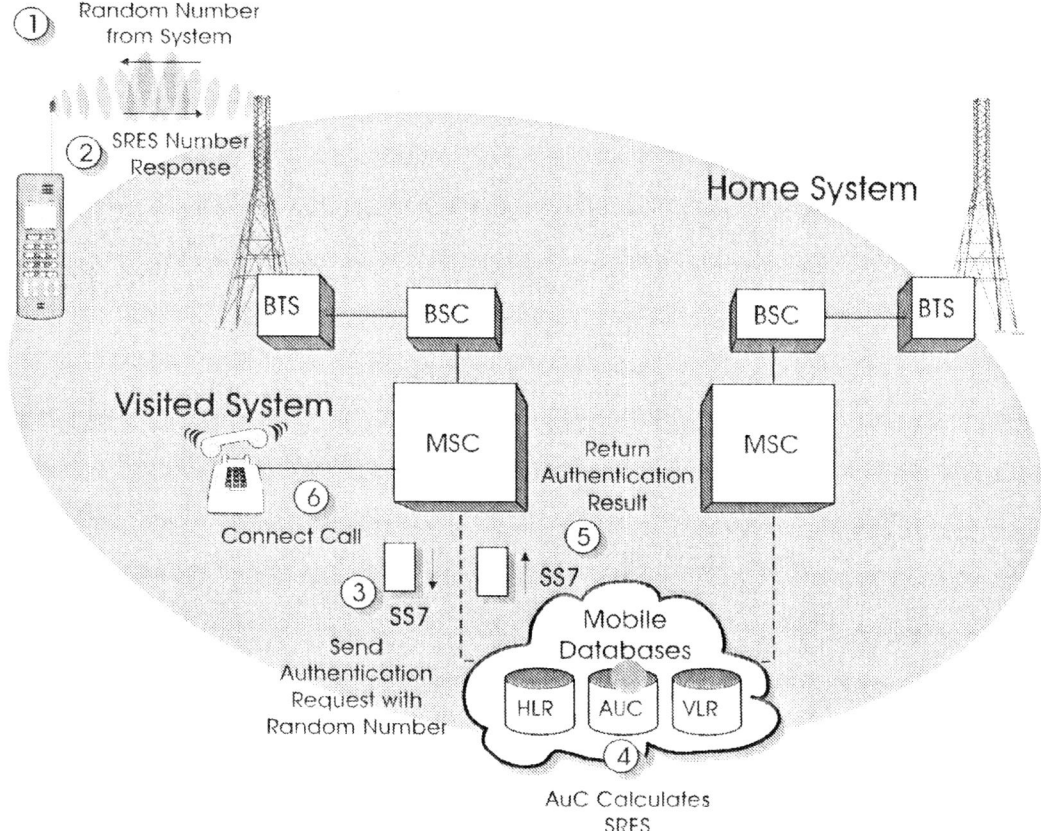

Figure 10.8. Mobile Authentication using SS7

SS7 Mobility Standardization and Interoperability

For universal operation and transparent access to user services by the mobile user, the industry has defined a number of specifications to be used by the service providers and the equipment suppliers. The major reference documents for Wireless SS7 and mobility features can be found in the following standards documentation.

ANSI-41 (American National Standards Institute) Revision C

ANSI TIA (Telecommunication Industry Association) TR45 (Technical Requirement 45)

ANSI TIA Wireless Intelligent Network (WIN) Interim Standard 771 (IS-771)

Telecordia Technologies (previously Bellcore) North American "Notes on the network"

European Telecommunications Standards Institute (ETSI) Global Service for Mobile Communications (GSM) document section 9.02

International Telecommunications Union (ITU) International Mobile Telephony 2000 (IMT-2000) specification

Chapter 11

Number Portability

Number portability allows an end-user of telecommunications services to retain their telephone number without impairment of quality, reliability or convenience, when changing any:

- Telephone Service Provider (Local Number Portability)
- Type of Service Being Offered (Service Number Portability)
- Location of Residence (Geographic Number Portability)

Telephone and Device Numbering

Each device within a network must have its own unique address. Some of the different types of addresses that are available include telephone numbers and data network addresses.

International Numbering Plan (ITU)

The International Telecommunications Union (ITU), a division of the United Nations, has defined a world numbering plan recommendation, "E.164." The E.164 numbering plan defines the use of a country code (CC), national destination code (NDC), and subscriber number (SN) for telephone numbering. The CC consists of one, two or three digits. The first digit identifies the world zone. The number of digits used for telephone numbers throughout the world varies. However, no portion of a telephone number can exceed 15 digits. There are several "E" series of ITU numbering recommendations that assist in providing unique identifying numbers for telephone devices around the world.

Figure 11.1 shows the world (telephone) numbering plan recommendation, "E.164" developed by the International Telecommunications Union (ITU). This diagram shows the numbering plan divides a telephone number into a country code (CC), national destination code (NDC), and subscriber number (SN) for telephone numbering. The CC consists of one, two or three digits and the first digit identifies the world zone. This diagram shows that the local number can be divided into an exchange code (end office switch identifier) and a port (or extension) code.

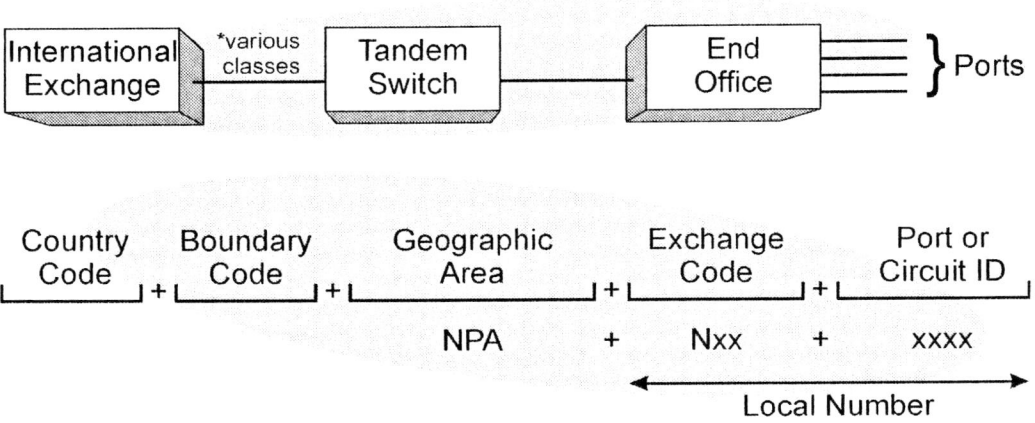

Figure 11.1., Telephone Numbering Systems

North American Numbering Plan (NANP)

The North American Numbering Plan (NANP) is an 11 digit-dialing plan that is used within North America. It contains 5 parts: international code, optional intersystem code (1 +), geographic numbering plan area (NPA), central office code (NXX), and station number (XXXX). The NPA code defines a geographic area for the serving telephone system (such as a city). The NXX defines a particular switch that is located within the telephone system. Finally, the station code identifies a particular line (station) that the switch provides service to.

Internet and Data Network Numbering

Most data network addresses are hierarchical where the beginning of the address identifies the entire network and each progressive address number (or group of numbers) identifies more specific parts within the network. Data networks are usually composed of several interconnected links. These links can be of different technologies with each of their end points identified by a unique numbering system.

Figure 11.2 shows how different types of data network addressing systems. This diagram shows a end-to-end data connection may transfer through many different networks and each network may use a different addressing system. This example shows that an end-user uses an Internet address to connect to a remote data device that has its own Internet address. The Internet address and its data is carried through the entire end-to-end communication in the data parts of each network packet. The first path connects the user to a company Ethernet network. The computers network interface card (NIC) has a 48 bit address unique to the Ethernet. Each packet that travels in the company's Ethernet network has it's own Ethernet address. Each packet of data from the end user includes the 32 bit Internet address. This packet (datagram) is encapsulated (stored) as part of the data message after the Ethernet address. The company's network is connected to an ISP by a high-speed frame relay connection. The frame relay access device (FRAD) has a unique identifier to the ISP. The ISP connects the data connection via asynchronous transfer mode (ATM) to the Application Service Provider (ASP).

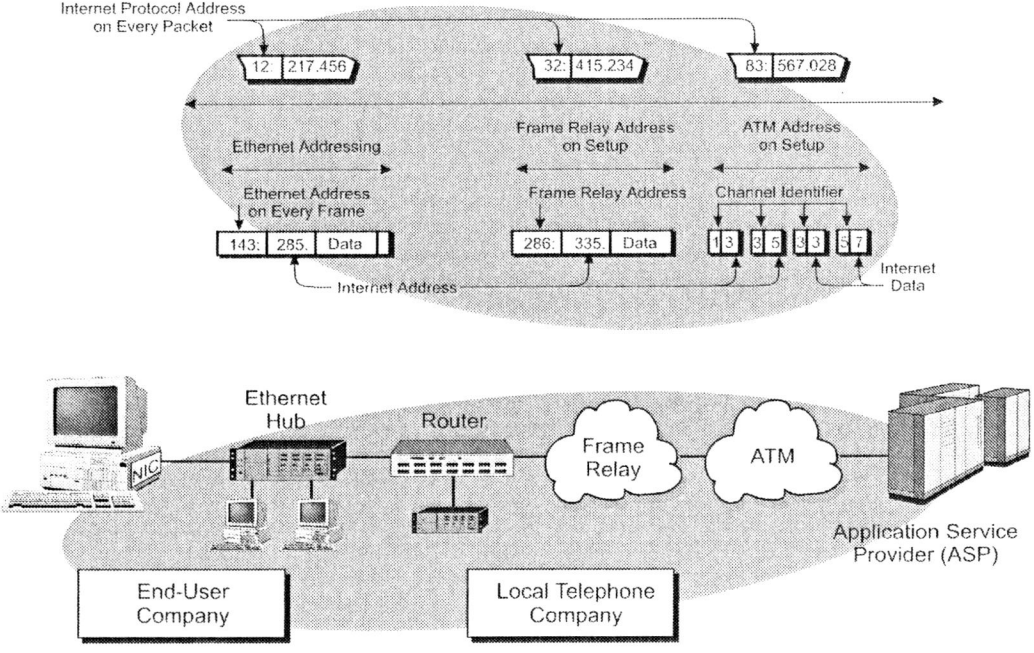

Figure 11.2., Internet and Network Numbering Systems

Number Portability

Number portability involves the ability for a telephone number to be transferred between different service providers. This allows customers to change service providers without having to change telephone numbers. Number portability involves three key elements: local number portability, service portability and geographic portability.

The need for number portability did not arise as a technical requirement for the network, rather number portability was mandated to make the telecommunications industry more competitive. The mandate of number portability reduces the barriers for new competitors to enter the marketplace. Because of number portability, new telecommunication service providers can offer competitive and/or advanced services to prospective customers without requiring a customer to change their telephone number.

Telecommunications Act of 1996

The Federal Communications Commission (FCC) mandated local Number Portability in the United States in 1996 to foster competition in the Local Exchange market. First implemented in 1997 by U.S. carriers, and expanded to be available in the Top 100 Metropolitan Statistical Areas (MSAs) and in any area within 6 months of the requested date.

The "porting" of telephone numbers as mandated by the Telecommunications Act of 1996 provides new "competitive" telephone service providers (telephone companies) the ability to take the existing telephone number away from the incumbent telephone company (usually well established with many customers) and maintain this phone number with the new service provider. Allowing the competitive telephone service provider to offer the same telephone number that the incumbent may have had for years is essential to the competitor getting business.

Local Number Portability (LNP)

The initial phase of number portability only addresses local number portability (LNP). LNP is the process that allows a subscriber to keep their telephone number when they change service providers in their same geographic area. Local number portability requires that carriers release their control of one of their assigned telephone numbers so customers can transfer to a competitive provider in the local area without having to change their telephone number. Providing LNP requires both the incumbent operators and competitive access companies to maintain databases of telephone numbers that have been ported. This allows the carriers to determine the destination of telephone calls that are delivered to the local service area.

The first part of the telephone number (NPA-NXX) usually identifies a specific geographic area and specific switch where the customer subscribes to telephone service. If a telephone number is assigned to another system (different NXX) in the same geographic area (same NPA), the interconnecting carriers (IXCs) connecting to that system must know which local system to route the calls based on the selected local service providers. In this case, the IXC must look up the local telephone number in a database (called a database dip) prior to delivering the call to the end customer.

Figure 11.3 shows an example of the typical operation of local number portability (LNP). In this diagram, a caller in Los Angeles is calling someone in Chicago who has kept (ported) their old phone number when they connected their service to a competitive local exchange carrier (CLEC). This required the incumbent local exchange carrier (ILEC) to move (port) the telephone number to a LNP database. The line connected to the customer from the CLEC actually has a new telephone number (which the customer is not likely to be aware of). The LNP database associates the new number with the old number. This example shows how the call can be routed from a LEC in Los Angeles to the new telephone line in Chicago using the old telephone number. The call is routed from Los Angeles, through a long distance provider (IXC) who knows by the dialed area code that it needs to connect the call into a local telephone company in Chicago. Because there are several local telephone service providers in Chicago, the IXC must look first into a LNP database to see if the number has been ported to a different service provider. This LNP database (ported telephone number list) must be available to the next to last switch (called "N-1") before the call reaches the end office switch. This LNP database search instructs the last switch to the actual number used for the final connection. The call is then routed to the correct local switching office (new line) so the call can be completed.

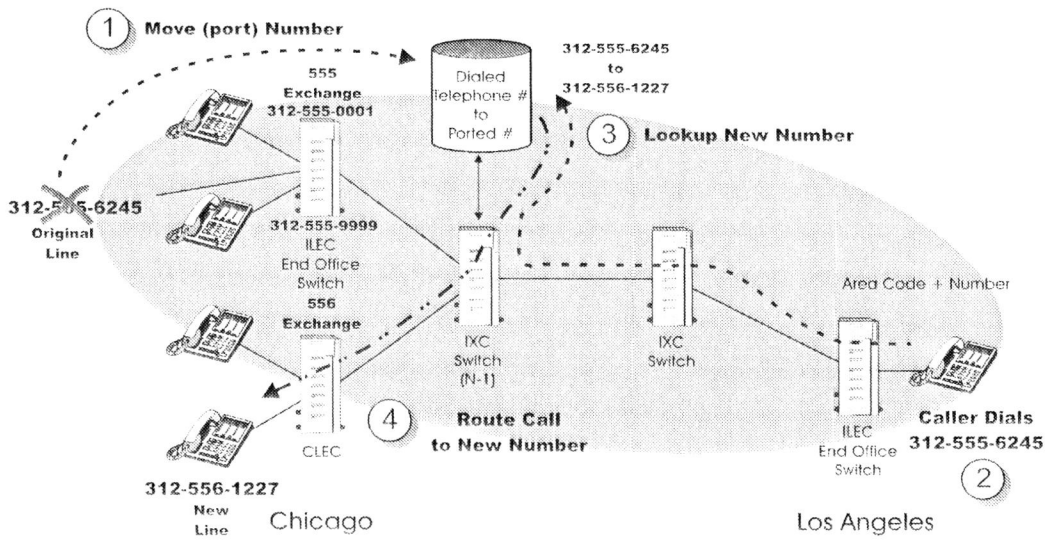

Figure 11.3.. Local Number Portability (LNP)

Service Number Portability

Service number portability allows a customer to take their telephone number to a different type of service provider. Service number portability involves determination of the type of service provider (e.g., wireless or wired) who is responsible for completing the call using the telephone number (e.g. area code and NXX.) Service number portability may differ from local number portability as the interconnection and call processing for different types of service providers may vary.

Wireless telephone number portability is an example of service number portability. Because of the complexities of call routing in mobile communication networks, wireless telephone number portability will be implemented at a later time. Until wireless porting is mandated in a specific area, a customer will need to change their mobile telephone phone number if they change their wireless carrier.

Geographic Number Portability

Geographic number portability involves the transfer of telephone numbers for telephone devices or services that are used outside the normal geographic boundaries of the service provider's original system or area. Geographic number portability allows a customer to keep their same area code when they move to new cities or other distant geographic regions. An example of an existing geographic number portability is 800 number or toll free services. However this service is implemented using the Intelligent Network, rather than the Location Routing Number method.

An example of geographic number portability, which is likely to be implemented ten years after local number portability, can be understood by this scenario: A family living in California decides to move to Florida and informs the broadband communications company in Florida they wish to retain their California area-code and telephone number!

SS7 Network Architecture to Support LNP

Implementing LNP has been one of the most complex modifications to the public switched telephone network (PSTN) to date. This immense challenge has been for phase one, where the end-user does not move location or type of service, rather "only" the service provider.

LNP impacts all the telecommunications networks, it requires all carriers, independent of underlying technology; to have the ability to appropriately route telephone calls in a ported environment. In the United States, the FCC has mandated that the next-to-last carrier (N-1) be responsible for changing their network architecture to support LNP, or pay on a per-call basis to have numbers routed by the default carrier.

This new LNP Network Architecture utilizes an SS7 routing method called the Location Routing Number (LRN) method. The LRN method will route a ported number to the switch on which the new ported number resides only after it determines it is a ported number. Determining this routing requires a query (search) for all the calls terminating to specific NPA-NXXs that have ported numbers. This method does not require any changes to the existing North American Numbering Plan (NANP), where all the area codes and exchanges have been defined for world-wide routing of calls.

LNP Network Elements

Service providers have specific LNP network elements (wireline or wireless) as well as "LNP-ready" software added to their existing network elements. In the network elements added (new) to support LNP are as follows:

- Number Portability Administration Center (NPAC)
- Service Order Activation (SOA) Center
- Local Service Management System (LSMS)

The existing network elements that require significant LNP-ready software are as follows:

- Billing and back-office systems
- Network Management and Local Operations systems
- Network Line Information DataBases (LIDB)
- Interoffice Trunking and SS7 Interface systems
- Signal Transfer Point (STP)
- Service Switching Point (SSP)
- Service Control Point (SCP)
- Emergency 911 Database (E-911)
- Local Exchange Routing Guide (LERG) Database
- Operator services systems

Call Flow to a Ported Telephone Number

The call flow to a ported telephone number involves digit analysis on the originated dialed digits to determine how to route the call. This analysis results in a search in the SCP database that determines the actual destination telephone number. The originating switch receives the Location Routing Number (LRN) from the SCP. The call is then routed to the destination switch based on the LRN.

Figure 11.4 shows an example of call flow in a system that has ported telephone numbers. In this example, the called party (User B) has changed telephone companies and desires to retain his original phone number. The process of routing the call occurs as follows:

1. User A (808-321-4567) dials User B (808-713-2222).

2. The Originating Switch performs digit analysis on the dialed digits to determine how to route the call. The Originating Switch determines that End-user B is in a portable NPA-NXX (808-713) and that End-user B does not reside on the Originating Switch.

3. The Originating Switch sends an SS7 message based on the dialed digits to the LNP SCP.

4. The LNP-SCP sends a message response containing the LRN (312-979-XXXX) of the Recipient Switch.

5. The Originating Switch receives the LNP SCP response and analyzes the data. The LRN is translated in the LNP Routing Tables and an Integrated Services User Part (ISUP) route from the Originating Switch to the Recipient Switch is determined. The LRN is stored in the Called Party address field and the dialed digits are stored in the Generic Address field of the ISUP Initial Address Message (IAM). The Forward Call Indicator (FCI) Translated Called Number Indicator is set to indicate an LNP message query has been performed.

6. The call is routed from the Originating Switch to the Recipient Switch based on the LRN.

7. The Recipient Switch receives and processes the contents of the IAM message. The switch determines that an LRN is received and that it matches its own LRN. The switch replaces the Called Party address parameter's contents with the dialed digits stored in the Generic Address field. The switch does a standard phone number analysis and locates the line associated with User B.

8. The Recipient Switch completes the call from End-user A to End-user B.

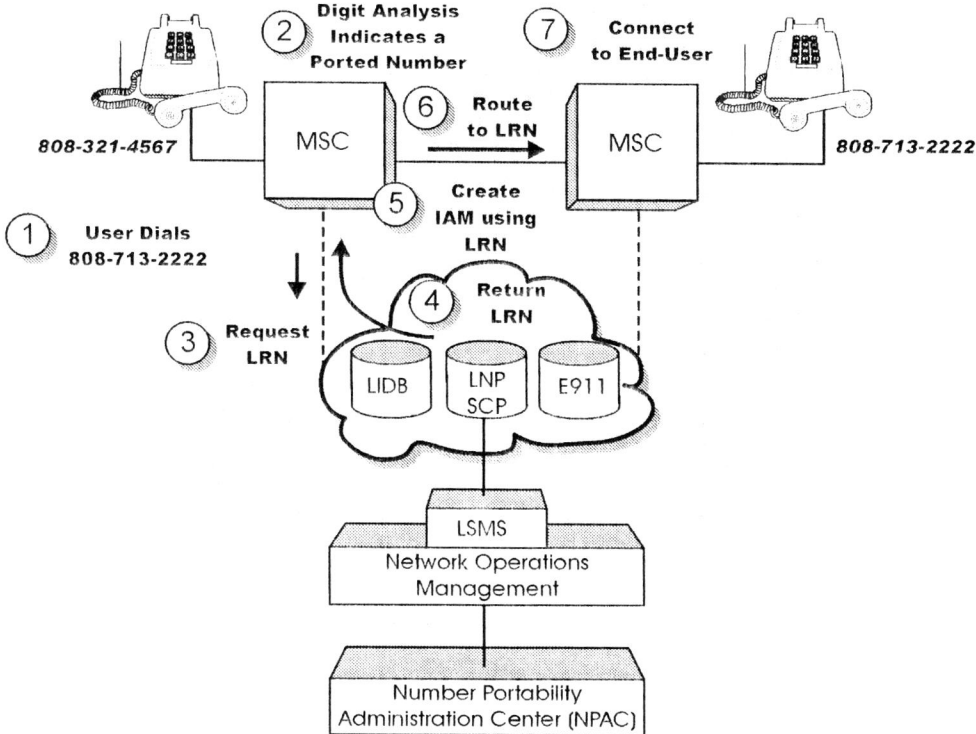

Figure 11.4. Call flow to a Ported Telephone Number

Administration of Ported Telephone Numbers

The telecommunications industry sought and selected a non-telecommunications company to administer the ported telephone number database. This "neutrality" was important, as LNP is a mandated ruling forcing competitive telecommunication carriers to share valuable information. This network element is called the Number Portability Administration Center (NPAC).

Lockheed Martin was selected and stabilized to be the administrator of the NPAC. In December of 1999 a separate company was formed by the same organization and is now known as NeuStar, Inc. NeuStar exclusively provides, reliable and responsive neutral third-party clearinghouse services to the telecommunications industry. The NPAC will serve to support:

- Telephone Service Provider (Telephone Company) Portability

- Location Portability (regardless of distance), and

- Service Portability

Administration duties in support of Local Number Portability involve many centralized activities; all performed by the NPAC. Consideration should be noted that LNP data administration is not to be confused with number administration. Number administration is the management of the entire telephone numbering resources where area codes and exchange codes are defined and managed at a 3, 6, or 10-digit level. Administration of LNP databases by the NPAC is done on a non-real time bases, whereby the following major tasks are performed:

Store a copy of the LNP database resident in participating networks. This is used as the master database for all national activities. For every ported telephone number, the NPAC LNP database stores:

- The network address (LRN) for call routing.

- The SS7 addresses for routing associated signaling messages to network databases, and

- The current local telephone service provider that "owns" the telephone number.

Coordinate service order processing between porting (competitive) service providers.

Propagate database updates (downloads) upon service order completion. Maintain NPAC Interface to Local Service Providers

Streamline NPAC services by directly connecting to carrier systems via mechanized (machine-to-machine) interfaces.

Offer, or interface to, Service Order Activation (SOA) systems

Offer, or interface to, Local Service Management Systems (LSMS)

Chapter 12

Broadband ISDN User Part (B-ISUP)

The Broadband ISDN User Part (B-ISUP) was developed to allow SS7 to support the control of voice and data communication on asynchronous transfer mode (ATM) systems. The ATM system is a connection-oriented packet switching system that allows for combined voice, data, and video services through the use of high-speed switches, prioritizing and transferring many small packets efficiently.

The key differences between ISUP and B-ISUP include new types of channel identifiers, several types of adaptation protocols, capability of assigning bandwidth allocations to switching systems, broadband and narrowband capabilities and how they interface with each other, and system recovery procedures for congestion control.

Figure 12.1 shows how the Broadband ISDN User Part (B-ISUP) functional part is located within the SS7 system. This diagram shows that the B-ISUP functional part is directly connected to the MTP3 layer and it is equivalent to the OSI network model application layers 3-7.

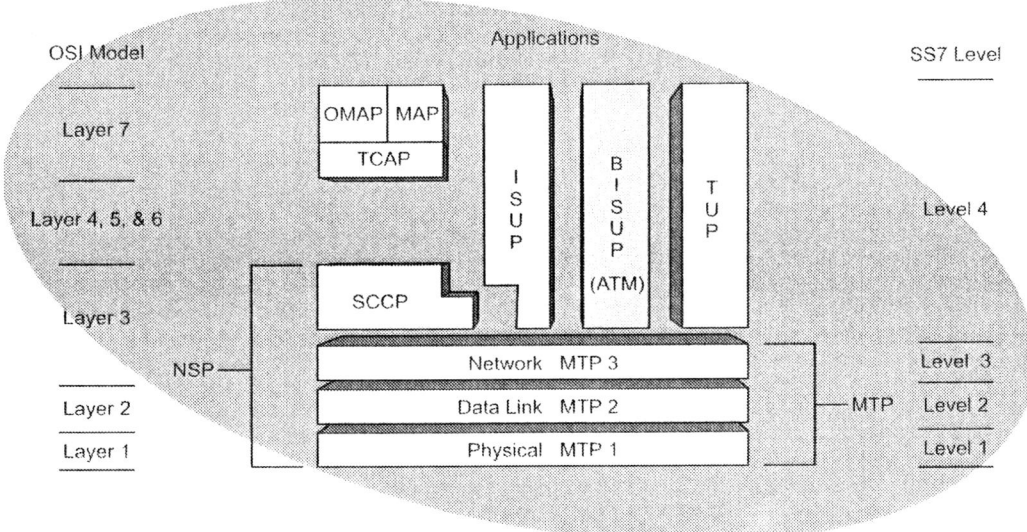

Figure 12.1, B-ISDN in SS7 Protocol Stack

ATM System

Asynchronous transfer mode (ATM) is a packet data transmission and switching system that transfers information by dividing all types of data into small fixed length packets of data (53 byte cells). The ATM system uses high-speed transmission (usually 155 Mbps or above) and is a connection-based system. When an ATM circuit is established, a path through multiple switches is setup and remains in place until the connection is completed. ATM service was developed to allow one communication medium (high-speed packet data) to provide for voice, data, and video service.

As of the 1990's, ATM has become a standard for high-speed digital backbone networks. ATM networks are widely used by large telecommunications service providers to interconnect their network systems. ATM aggregators operate networks that consolidate data traffic from multiple feeders to transport different types of media (voice, data, and video).

The ATM switch rapidly transfers and routes packets to the pre-designated destinations. To transfer packets to their destination, each ATM switch maintains a database (called a routing table). The routing table instructs the ATM switch to which channel to transfer the incoming packet to and what priority should be given to the packet. The routing table is updated each time a connection is setup and disconnected. This allows the ATM switch to forward packets to the next ATM switch or destination point without spending much processing time.

The ATM switch also may prioritize or discard packets that it receives based on network availability (congestion). The ATM switch determines the prioritization and discard options by the type of channels and packets within the channels that are being switched by the ATM switch.

Figure 12.2, shows a functional diagram of an ATM packet switching system. This diagram shows that there are three signal sources going through an ATM network to different destinations. The audio signal source (signal 1) is a 64 kbps voice circuit. The data from the voice circuit is divided into short packets and sent to the ATM switch 1. ATM switch 1 looks in its routing table and determines the packet is destined for ATM switch 4 and ATM switch 4 adapts (slows down the transmission speed) and routes it to it destination voice circuit. The routing from ATM switch 1 to ATM switch 4 is accomplished by assigning the ATM packet a virtual circuit identifier (VCI) that ATM switch can understand (the packet routing address). This VCI code remains for the duration of the communication. The second signal source is a 384 kbps Internet session. ATM switch 1 determines the destination of these packets is ATM switch 4 through ATM switch 3. The third signal source is a 1 Mbps digital video signal from a digital video camera. ATM switch 1 determines this signal is destined for ATM switch 4 for a digital television. In this case, the communication path is through ATM switches 1, 2, and 4.

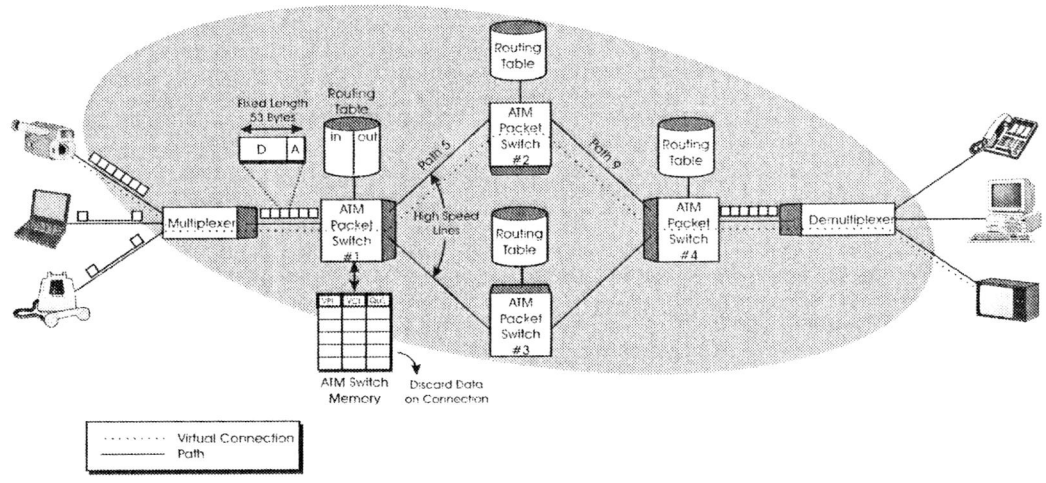

Figure 12.2, ATM System

Broadband Parameters

Broadband communication sessions have additional parameters that may be setup and maintained by the SS7 system as compared to the standard ISDN user part (ISDN-UP.) These include new path and channel identifiers, different types of protocol adaptation layers, cell rate (desired bandwidth,) broadband bearer capabilities, interfaces to narrowband channels (potential limitations,) and thresholds for congestion recovery procedures.

New Channel Identification Parameters

ATM system assigns a circuit by 2 identifiers; virtual path identifier (VPI) and a virtual circuit identifier (VCI). These parameters must be sent as new fields along with the addressing information normally used by the SS7 messages.

ATM Adaptation Layer (AAL) Parameters

The ATM adaptation layer (AAL) is a set of standard protocols that translate (adapt) user traffic into a size and format that can be contained in the small payload of an ATM cell. After the data is divided, transmitted, and received, is recombined to its original form at the destination. This process is called segmentation and reassembly.

Because some services require different types of communication characteristic parameters (e.g. real time voice communication compared to file data transmission,) different types of protocol conversions are used to allow for different services.

All AAL protocol adaptation functions occur at the ATM end-station rather than at the switch. These protocol layers are designed to allow for constant bit rate (CBR), variable bit rate (VBR), unspecified bit rate (UBR), and other types of services.

Cell Rate

The cell rate is the number of cells per second (53 byte ATM packets) that are required to be passed through a switch for a particular call or communication session. Because the switching capacity of an ATM switch can dynamically change, the bandwidth of the link can be assigned by programming the cell rate.

Broadband Bearer Capability

Broadband bearer capability is the ability of the bearer to provide the necessary bandwidth that is needed by the application for the bearer service. B-ISUP can define the broadband bearer capability by indicating timing requirements, continuous bit rate, or variable bit rate traffic.

Narrowband Inter-working Capability

Narrowband inter-working capability defines the narrowband connection that may be attached to the B-ISUP connection. This narrowband connection may limit the end-to-end bearer service capability. It may have full ISUP signaling available at the end of the connection or it may be a narrowband connection with limited signaling and control capability.

Congestion Threshold Level

Because the ATM system uses connection-oriented packet switching and the cell rates of each connection can vary, it is possible (and likely) that some ATM switches will become congested. The possibility of congestion requires the system to dynamically change as switches become congested. To provide for better congestion control, B-ISUP signaling allows for the programming of congestion threshold levels. This allows the system and applications to better recover from congestion conditions.

Congestion is indicated between switches by flags contained within the ATM cells. An explicit forward congestion indication (EFCI) is a field contained within an ATM cell header that is used to indicate to other network elements (e.g. switches) that a switch is experiencing or will likely experience a congested state. The use of EFCI information allows other switches or systems to adapt or lower their cell rates.

Chapter 13

Intelligent Networks

Intelligent networks are communication systems that have the ability to process call control and related functions via distributed network transfer points and control centers as opposed to a concentrated in switching system. The intelligence of the telecommunications network refers to the complexities involved in deciding if and how a telephone call or a data connection shall be routed.

Intelligent networks provide uses with the ability to manage call processing intelligence. An intelligent network could be instructed to proceed with or change its call processing based on information it learns about users of the network. For example, the telecom network today could discover that for the last ten years, each and every Friday at 4 p.m. Farmer Brown can be reached down at the Post Office until 10:00 p.m. using this information; the telecommunication network could automatically redirect the calls for Farmer Brown.

Migration of Intelligence in the Network

The Intelligent Network (IN) has evolved from localized intelligence in the telephone switch. Telephone switches simply connect ports or communication circuits to each other. Network intelligence is used to control the switching systems using a signaling protocol such as Signaling System 7

(SS7). Intelligent network standards define the processes used to extract the functionality out of the switching systems and enhance by controlling the call via the SS7 network.

The benefit of migrating to an intelligent network is the ability to improve on existing services and develop new sources of revenue. The migration is done with the following objectives:

> To introduce new services rapidly and modify existing services without switch intervention.
>
> To provide unique (per user) service customization rapidly and efficiently; in some cases allowing the subscriber, to control of their own services.
>
> To maintain vendor independence, creating a more competitive environment that implements more quickly and inexpensively.
>
> To establish open interfaces allowing service providers to introduce network functions quickly while still maintaining stringent network operations standards.

IN technology allows the continued use of the embedded base of stored program-controlled switching systems by separating the service-specific functions and data from other network resources. This reduces the dependency on the switching systems for custom (unique) software programs and it may allow other vendors or even the operators of the systems to create and customize advanced services.

In an IN, a network node known as the Service Control Point (SCP) contains programmable service-independent capabilities (also called service logic) that are controlled by the service provider. The SCP also contains service-specific data that allows service providers and their customers to customize services. The IN allows for unified/centralized services, as well as unique services that are customized to meet even a single user's need.

Because the service logic is removed from the switching infrastructure, network providers can offer market-focused service trials by loading test logic in an SCP and triggering only those calls in the trial. The migration to an intelligent network requires that the Service Switching Point (SSPs) in the network, surrender control of the call to the "higher" intelligence of the SCP. Until the entire network switch fabric has been replaced with completely "dumb" switches, calls will be handled by a combination of switch-based services, and intelligent network based features.

The ability to open the basic switching process to external influence is done by defining a call model, called the Basic Call State Model. This model describes the activities required to establish and maintain communication paths for a telephone call. Furthermore, the basic call model identifies points in the basic call where triggers can occur.

For Intelligent Network (IN) features, the SSP surrenders control of the call at trigger checkpoints, to the network Service Control Point (SCP). The earliest example of this type of IN call control can be found in an 800 or toll free calling scenario.

> When an 800 or toll free number is dialed, the switch pauses, and immediately requests the SS7 subsystem to query an enhanced 800 service control point (SCP) database for translation of the 800/toll free number. The translation is to a switch specific telephone number containing an area code, an exchange code, and line number used for traditional routing in the network.

Figure 13.1 shows the layers of intelligence in the public telephone network as they are split to perform network-wide enhanced services. The Service Switching Point (SSP) is the lowest level of intelligence, whereby it mechanizes the connection points of a call as directed by the service logic in the Service Control Point (SCP). The SSP receives an initial indication (such as the phone taken off-hook) that a call is to be placed by this particular line. In a scenario where a line is identified as a "hot-line" such as the telephones found on a rental car kiosk, the SSP immediately sets a trigger (or triggers) indicating to the SS7 subsystem to take control of the routing of the call.

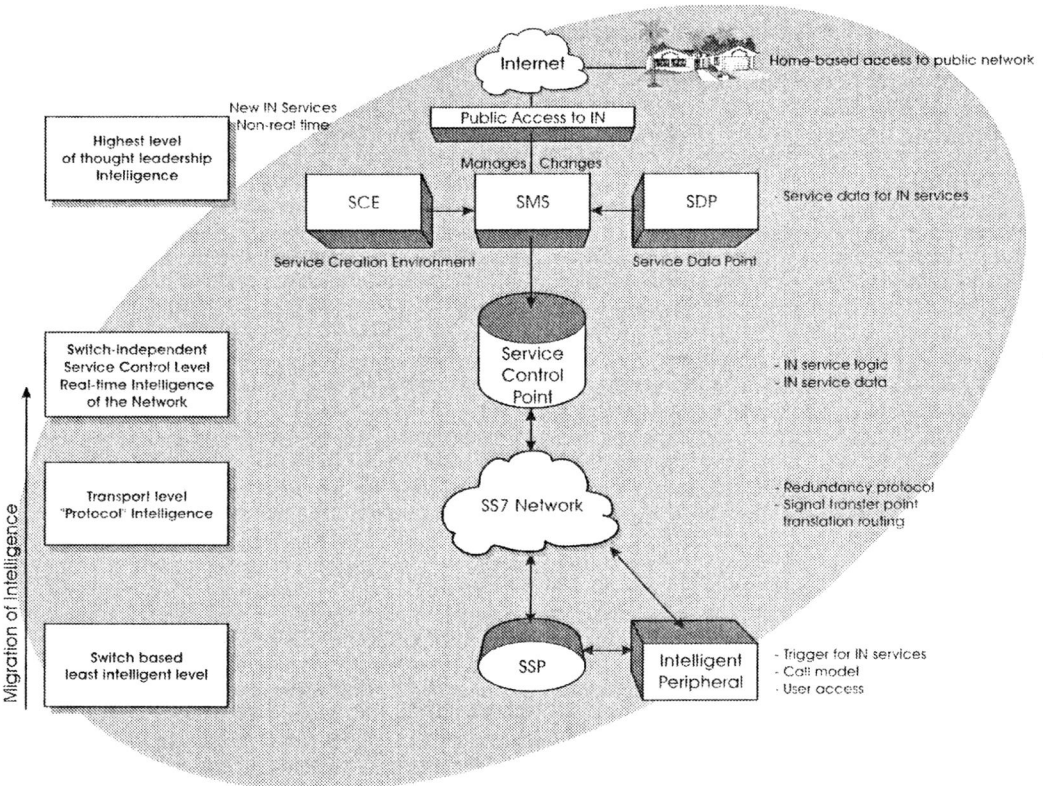

Figure 13.1. Migration of Intelligence in the Network

The SS7 subsystem of the SSP then hands off the query-for-call-routing-instructions request to the SS7 Network (the cloud in Figure 13.1) for transport to the appropriate network intelligence database, such as an 800 or toll free application within a Service Control Pont (SCP). The protocol intelligence of the SS7 network allows for complete redundancy while accessing SS7 network nodes. The SS7 network protocol also provides for global-title-translation by network Signal Transfer Points (not shown, but is within the SS7 network cloud) to route the SS7 query message to the correct network

SCP and application requested by the SSP. SCP returns to the SSP the data that the SSP needs to proceed with call processing.

Real-time transactions (or queries) occur at the network SCP. During the initial call processing phase, the originating switch (SSP) momentarily pauses the call in progress, it awaits a response from the SCP. This critical time period is when the Service Control Logic (in the SCP) manipulates the handling of the call based on IN Service Data uniquely applied for this IN call. In our rental car scenario, the IN Service Data would be time-of-day, day-of-week, geographic location, and the selections of destination phone numbers of the car rental's national customer service agents; all this immediately after the customer picks up the "hot-line" telephone.

In real-time IN transactions may also occur at the distant end, just prior to the call being terminated (ringing the phone). These types of services use terminating "triggers" where the call is paused by the terminating SSP awaiting instructions from the intelligent network. One example of this type of service (there are many) can be described as a Follow-me Forwarding Service; where the terminating control logic first seeks the IN Service Data for time-of-day, day-of-week, Calling Party Line Identification, and related forwarding number before (or if) forwarding the call.

Service Creation and Management

The uppermost level of network intelligence as seen in Figure 13.1 is the thought leadership concept of Service Creation. For operational management of new and changed services and updates to the IN Service Data a stand-alone, non-realtime Service Management System (SMS) is architected into the network.

The service-independent switch architecture allows the provision and operation of new services to be rapidly and efficiently managed through the SMS. This allows services to be added or changed without having to redesign switching equipment or populate new software in every switch.

Service Creation Environment (SCE)

Creation Environment (SCE) is a development tool that is used for the deployment of Intelligent Network services. SCE tools enable developers to prototype, test, and roll out new enhanced telephony applications quickly in response to growing market opportunities. This is customarily accomplished by a development environment that is a set of graphical tools for designing specifically to telecom network applications. These graphical tools manipulate a combination of telecom specific, programming building blocks known as Service Independent Building-blocks (SIBs) used to create service flows for the SCP.

Creating services is a matter of graphically combining and arranging these building blocks to form "Service Logic" for the SCP. The service creator need not be skilled in any programming language. The graphical service logic contains the services flows and data points for execution of the service in real-time when the SSP calls upon this service.

Figure 13.2 displays an approach of creating service logic by linking specific telecom Service Independent building Blocks (SIBs) in accordance with a service requirement submitted by product marketing. This service logic (chain) starts at a Point of Initiation (POI) and ends at a Point of Return (POR).

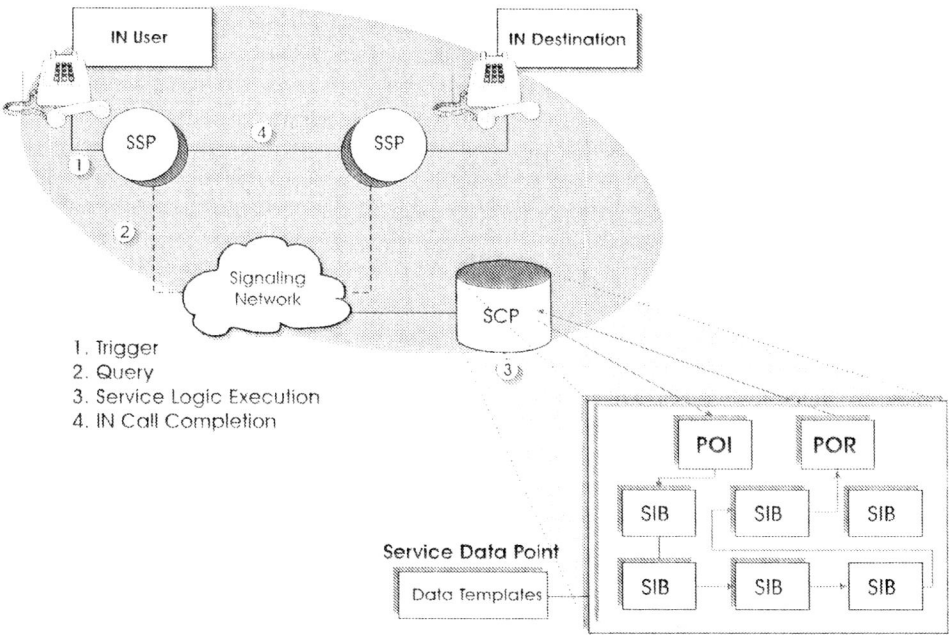

1. Trigger
2. Query
3. Service Logic Execution
4. IN Call Completion

Service Data Point

Data Templates

Mirrors-Service Creation Environment

Figure 13.2. Service Logic using SIBs

The SIBs are specialized call processing SIBs that represent the interaction between the SSP and the SCP. Table 13.1 typifies the purpose of telecom SIBs and their corresponding point-of-initiation (POI), and point-of-return (POR) within a call flow.

	SIB	POI	POR
Data Handling	Algorithm, Compare, Distribution, Limit, Screen, Service, Data, Management, Translate, Verify, Charge	Call Originated, Address Collected, Address Analyzed, Prepare to Complete Call, Busy, No Answer, Call Acceptance, Active State, End of Call	Continue with Existing Data, Proceed with new Data, Handle as Transit, Clear Call, Enable Call Party Handling, Initiate Call
Call Control	Queue, User interaction		
Collection of Information about a Call	Log Call Information		
Monitoring of Network Resources	Status Notification		

Table 13.1. SIB Purpose and Place in Call

A Service Data Point (SDP) provides all the types of data that can be used in service control logic. The service data is encapsulated in data templates as defined by the service creator. These data templates contain call variables such as integers, strings, or even database tables, and all can be easily created within the Service Creation Environment (SCE). The service creator is also able to define where to obtain the data during real-time operation, for example; on the SCP platform itself or in a remote database located elsewhere on the SS7 Network.

Service creation efforts likely start with a documented service development process and a laboratory containing all equipment and tools required to test the services and their interaction in the switched network. This laboratory environment should include SSPs, SCPs, IPs, and STPs. The bad news regarding centralized call-control is that major network and customer service disruptions can occur if a Service Creation design fails to operate as planned.

Interoperability of interconnecting IN services across different service providers must be taken into consideration. This includes interoperability across various SSP products and across different SCPs and STPs for the current network. As the information infrastructure evolves, there will be more interaction between the different service provider networks. Network reliability in this interactive network is a challenge.

Service Management System (SMS)

To deliver service control logic and manage service data, an administrative system must be employed. The Service Management System (SMS) operation may range from a simple Web-based graphical user interface, to a complex management application system that monitors the entire status of all intelligent network related resource groups so administrators can quickly and easily obtain detail needed to make service management decisions.

The Service Management System (SMS) is operated under non real-time conditions. This provides the opportunity to test and retest service components under various scenarios combined with service data, before downloading to the real-time SCP.

Subscriber access to the Intelligent Network is buffered by the Service Management System. An Internet provisioning interface may be provided for Internet-based access to service and subscriber data. It is likely that only a few sophisticated subscribers, or large corporate entities are able to concur with the stringent Applications Programmers Interface (API) requirements to the Service Management System. That said, user self-service and management will be essential to the success of the next generation intelligent network.

Wireless Intelligent Networks (WIN)

The wireless (cellular) population has grown at an unprecedented rate since its inception in the early 1980s, surpassing all predictions. As supply and demand would warrant, subscribers have become very demanding, to a point where they expect their wireless device to perform as if it were wired. Part of the solution is to provide radio coverage absolutely everywhere, and the other part of the solution, meeting subscriber's "wireline like" expectations is in the enhancement of the systems that would provide wireline transparent services. This is being accomplished by a steady migration to an Intelligent Network (IN) services architecture.

This intelligent network migration allows the service providers to quickly develop and introduce enhanced "wireline like" services across their wireless network without the arduous task of major upgrades to the switching elements similar the landline tasks. In addition to creating revenue earning opportunities for the wireless network operator, this migration provides the subscribers with a high level of personal control over their services. In fact the majority of wireless carriers have marketed their product as "Personal Communication Services" (PCS) leveraging on their addition of new radio frequencies, but more importantly the intelligence of the network. This network intelligence provides a value proposition to the subscriber significant enough for many subscribers to actually replace their wired phone service with wireless service.

IN-based services provide for an important continuity of services across network providers, a feature unheard of years ago. Co-opetition (competitors cooperating) to provide wireless subscribers seamless access to wireless intelligent network (WIN) services.

Wireless Intelligent Network (WIN) technology enables service providers to offer subscribers capabilities that blend both wireless and wireline control of their communications. This emerges as a "One Number" service for example, where users have tethered and un-tethered services with one voice mail box, one phone number and one bill.

New wireless intelligent network-based (WIN) services will potentially allow all end-users to harness the power of their mobility in ways that suit their business and personal lifestyles. To appreciate the potential of WIN services, a WIN-based call scenario is provided here:

Figure 13.3 show a Calling Party Controlled Completion (CPCC) service example that is a WIN feature. In this example, a calling party is asked to pay for a (normally free) call destined for a wireless phone. The scenario is such that the wireless phone user is roaming in another market that would incur long distance charges and possibly roaming charges. The wireless phone user subscribes to the CPCC WIN feature to control the costs of his calls while traveling. The WIN subscriber has listed the few VIP numbers willing to call for free as usual. All other callers are routed to a network-based Intelligent Peripheral that calculates the cost of the call, and provides a method for the calling party to pay, if they decide they must talk to the wireless phone user.

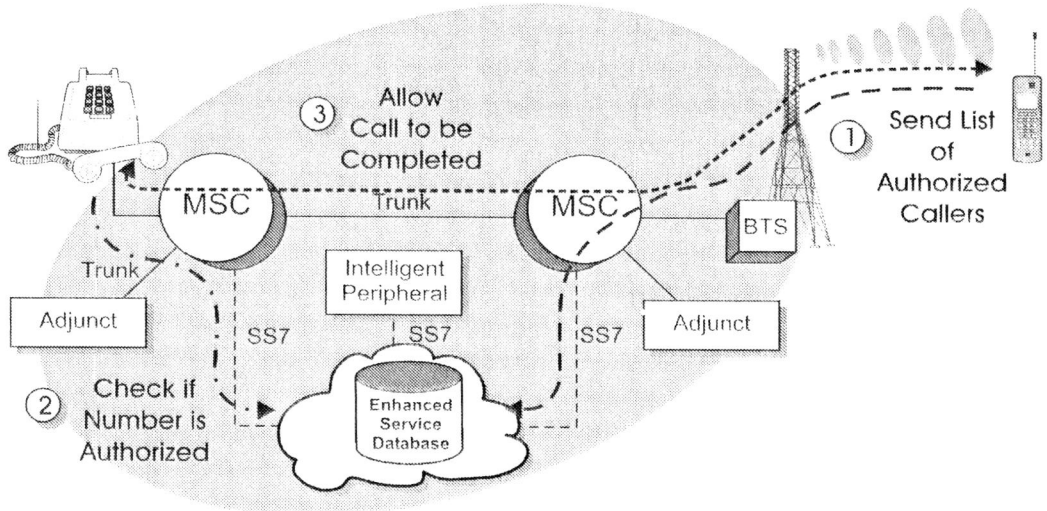

Fig 13.3. Calling Party Controlled Completion (CPCC) Architecture

As some users have relinquished their wireline phone altogether, a WIN service providing a "dual-profile" service would enable subscribers to set up a business profile and a personal profile for their wireless phone. The service logic for this type of segregation could be a simple or as complex as the user desires. The service creation method is identical to the service creation environment (SCE) used in the landline description above.

Customized Application of Mobile Enhanced Logic (CAMEL)

CAMEL is an example of a European initiative much like the WIN architecture. CAMEL provides similar WIN functionality such as prepaid service, calling party pays (especially across international boundaries) and screening applications. It supports roaming agreements between operators and protocols, a concept once thought impossible. One advantage CAMEL has is the ubiquitous nature of the Global System for Mobile Communications

(GSM) networks standardized by most world operators. Virtual Private Network (VPN) services have become increasingly important for GSM mobile users. Because of their global markets, operators are more willing to allow their subscribers to create and manage their own enhanced wireless services. And to travel across the continent using your GSM phone might normally incur charges from several countries before you get to your destination; so a "one bill" application of personal mobility is desirable.

European Telecommunications Standards Institute (ETSI) is the body providing telecom standardization and integration of these value added services. Implementing CAMEL can be regarded as the introduction of IN into GSM networks. ETSI has now handed over responsibility for GSM standards to the Third Generation Partnership Project (3GPP).

Chapter 14

SS7 and Internet Protocol (IP)

Public telephone systems today use the Internet for both; telephony voice and to carry signaling system 7 (SS7) messages. The Internet can provide reliable communication services by using the packet based transmission technologies used by IP-based protocols.

There is a difference between the Internet and IP based networks. IP based networks uses Internet protocol to route information within the network. The Internet is a public data network that interconnects private and government computers together. An IP based network does not have to be part of the Internet and it is possible for an Internet network operator to partition their data network to allow for different quality of service (QoS) levels. As a result, it is possible to reliably send SS7 control messages over IP based networks that may be part of the public Internet.

The Internet transfers data from point to point by packets that use Internet protocol (IP). Each transmitted packet in the Internet finds its way through the network switching through nodes (computers). Each node in the Internet forwards received packets to another location (another node) that is closer to its destination. Each node contains routing tables that provide packet forwarding information. These routing tables may be dynamically changed as a result of new connections or paths that may become available through the network. This is different than the SS7 system that allows the operator to have more precise control over the routing tables.

The use of IP based networks for voice, data, multimedia, and signaling offers new potential levels of network efficiency (utilization). A key protocol that has been developed to allow the sending of signaling control messages over IP based networks is Signaling Transport (SIGTRAN). The SIGTRAN protocol stack utilizes the Stream Control Transmission protocol (SCTP). SCTP is an IP protocol that combines near-real time data transfer with reliable packet delivery and validation. To allow the use of protocols with SS7 systems, several protocol adaptation layers have been created. These protocols adapt the message structures and flow of messages to emulate the message transfer parts (MTP) of the SS7 protocol stack.

Internet telephone systems are primarily composed of media gateways (MGs) and one or more media gateway controllers (MGCs). When Internet telephone systems interconnect to other networks such as the Public Switched Telephone Network (PSTN), they use signaling gateways (SG) or Network Gateways (NGWs). Some of the more common IP Telephony Systems include session initiation protocol (SIP), media gateway control protocol (MGCP), MEGACO, and ITU's H.323 protocol.

SS7 and Internet Protocol (IP) Signaling Systems

SS7 messages can be directly transported over IP networks or the functional equivalent of SS7 control message can be sent as control messages (e.g. text based messages) directly between elements connected to a data network (e.g. the Internet).

Figure 14.1 shows that SS7 signaling systems can be interconnected with voice over data networks and that SS7 messages can be transported over the Internet protocol. This diagram shows that analog and digital telephones are connected to the PSTN. To interconnect these telephones to voice over data network telephones, the media portion of each communication session is routed through a media gateway where it is converted from the PSTN circuit switched form to a IP packet data media format (packetized voice.) This diagram shows that the packet media can be routed through a data network (e.g. Internet) to an endpoint communication terminal such as a multimedia computer or an IP telephone. This diagram also shows that the SS7 network can control the PSTN through SS7 signaling messages and it can communicate to the media gateway through IP signaling messages.

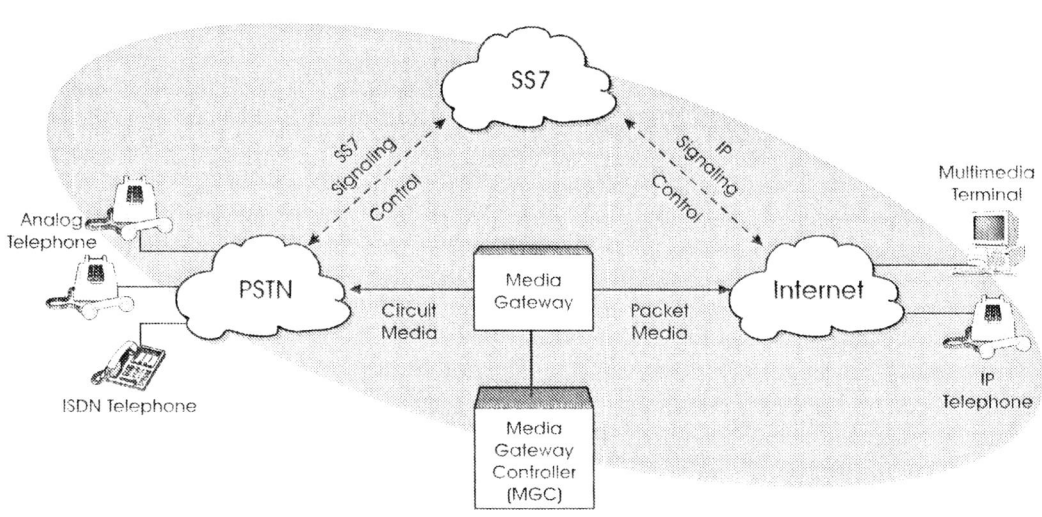

Figure 14.1. Hybrid SS7 and Internet Protocol Network

Figure 14.2 shows a basic voice over IP protocol stack. This diagram shows that the layers for a Voice over IP (VoIP) system may be composed of multiple technologies. The physical layer may be dial-up modems, DSL, cable modems, or any other physical transport system that can transfer digital information. The data link layer includes point-to-point protocol (PPP) and other link management systems. The network layer is the Internet protocol (IP). The session layer uses H.323 or SIP to setup, coordinate and teardown communication sessions. The presentation layer is the conversion of raw information into a usable form such as G.729 and G.711 speech coding. The application layer may include voice, data, and video media display and control.

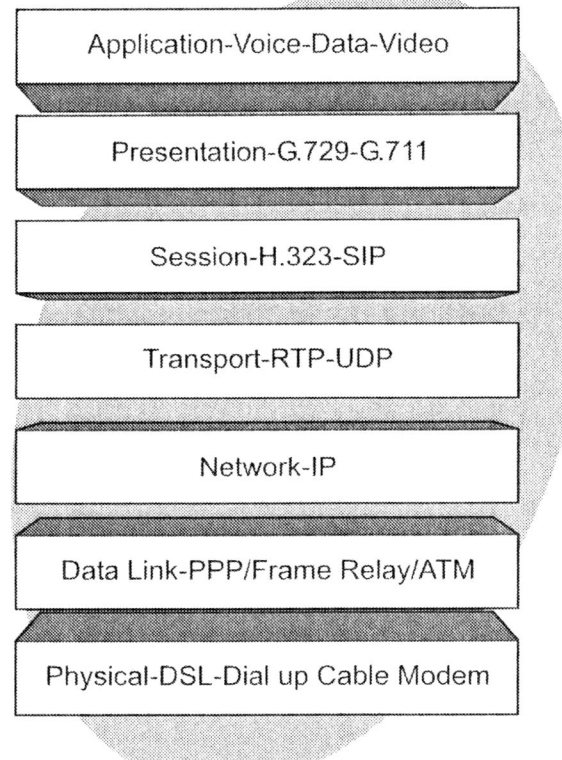

Figure 14.2. VoIP Protocol Stack

Signaling Transport (SIGTRAN)

A set of standard that were defined by the Internet engineering task force (IETF) that contain a set of protocols that are suitable to provide signaling control messages (such as SS7 message) over in Internet Protocol (IP) network.

Figure 14.3 shows that the Sigtran protocol stack is composed of the packet transport layer (IP), common signaling transport layer, and adaptation protocol layer. This protocol stack allows the Sigtran system to transport signaling control messages on a connectionless IP based system. The IP communication channel is managed by the SCTP connection-oriented protocol layer to allow for sequential and secure transport. The adaptation layers convert the protocols (e.g. IP addressing to Point Code addressing) between the Sigtran system and the SS7 system.

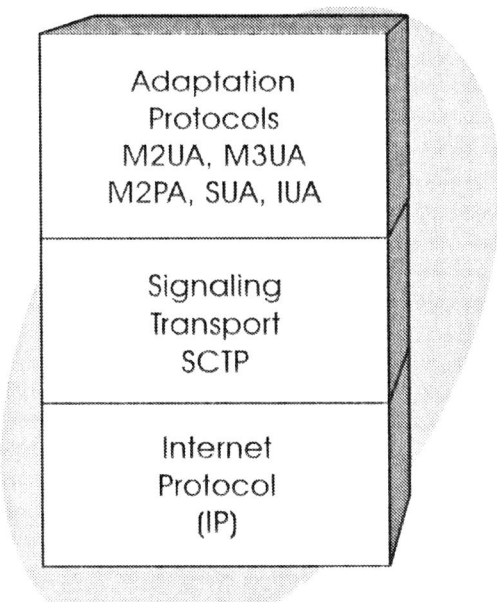

Figure 14.3, Sigtran Protocol Stack

Signaling Gateway (SG)

A signaling gateway (SG) is used to interface a signaling control system (e.g. such as SS7) and a network device (e.g. a transfer point, database, or other type of signaling system). The signaling gateway may convert message formats, translate addresses, and allow different signaling protocols to interact.

Media Gateway (MG)

The media gateway interfaces the PSTN to multimedia data communication systems as specified by MGCP. The media gateway is responsible to interface the different types of media formats between the public and data networks.

Figure 14.4 shows the functional structure of a media gateway (MG) device. This diagram shows that this gateway interfaces between a public telephone network line side analog connection to an Internet packet (IP) data network connection. The overall operation of the voice gateway is controlled by a media gateway controller (MGC.) The MGC section receives and inserts signaling control messages from the input (telephone line) and output (data port). The MGC section may use separate communication channels (out-of-band) to coordinate call setup and disconnection.

Signals from the public telephone network pass through a line card to adapt the information for use within the media gateway. This line card separates (extracts) and combines (inserts) control signals from the input line from the audio signal. Because this audio signal is in analog form (another option could be an ISDN digital line side connection,) the media gateway converts the audio signal to digital form using an analog to digital converter. The digital audio signal is then passed through a data compression (speech coding) device so the data rate is reduced for more efficient communication. This diagram shows that there are several speech coder options to select from. The selection of the speech coder is negotiated on call setup based on preferences and communication capability of this media gateway and the media gateway it is communicating with. After the speech signal is compressed, the digital signal is formatted for the protocol that is used for data communication (IP packet.) This diagram shows that the call processing section of the media gateway is not part of the gateway. It is a separate controller that commands the gateway to insert messages in the media stream (in-band signaling) or it may communicate with the other gateway through another media gateway controller (MGC.)

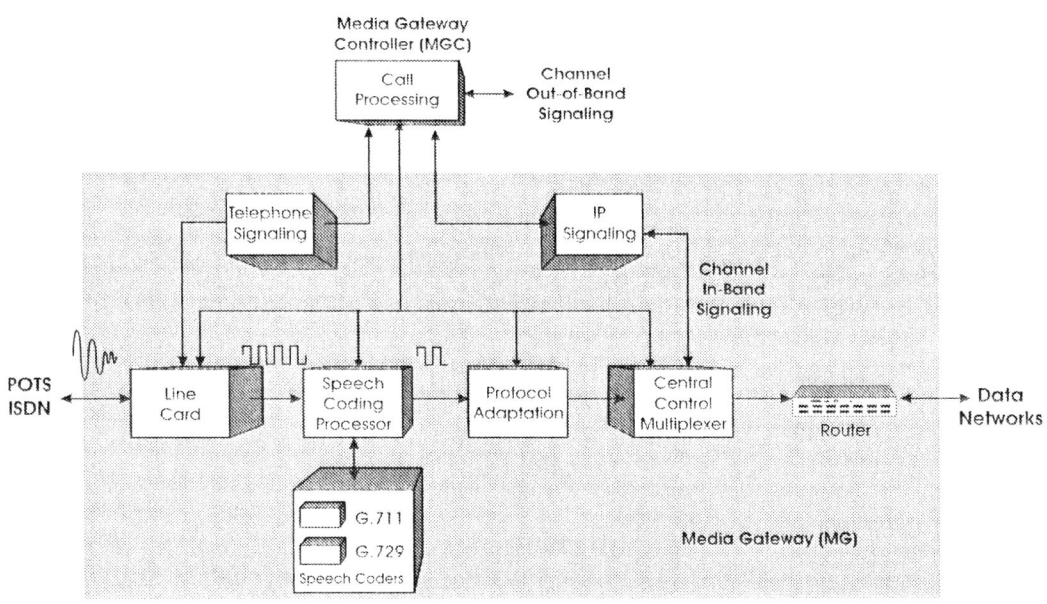

Figure 14.4. Media Gateway (MG)

Media Gateway Controller (MGC)

The media gateway controller is the call processing portion of a PSTN gateway that acts as a surrogate call management system (CMS). The MGC controls the signaling gateway and the media gateway (MG). The protocols between the MGC and MG include the media gateway control protocol (MGCP), the IETF/ITU Media Gateway Control (MEGACO)/H.248 protocol, and ITU's H.323. The MGC acts as a call agent coordinating sessions between devices. Signaling between MGCs (agents) may use SIP or H.323 protocol.

Stream Control Transmission Protocol (SCTP)

Stream control transmission protocol (SCTP) is an Internet protocol that is used to coordinate the sending of near real time information (e.g. signaling control) over packet based communication systems. SCTP protocol was developed to offer short transmission delay time and reliable data delivery. In essence, SCTP combines the benefits of efficient user datagram protocol (UDP) and reliable transmission control protocol (TCP).

Figure 14.5 the basic protocol stack for the SCTP data stream transmission protocol. This diagram shows that the SCTP protocol stack is responsible for maintaining a path through an IP (simulated connection oriented) network. Each SCTP packet includes information that allows the validation of each packet. The SCTP protocol allows the combination of several chunks (data streams or messages) of data into each packet. The SCTP system includes information that allows packets to be routed to avoid congestion as delays can cause serious challenges with signaling and streaming applications. The SCTP packets can be divided as they pass through networks that have maximum packet sizes than are originated by the SCTP user. The SCTP system contains information to allow the sequencing of packets.

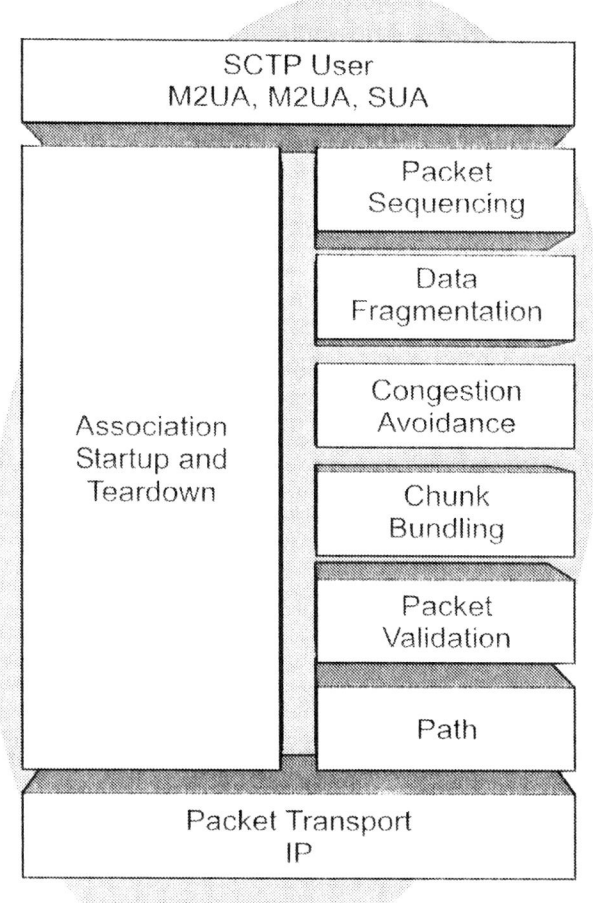

Figure 14.5. Protocol Stack

Stream control transmission protocol (SCTP) packets contain a common header and variable length blocks (chunks) of data. The SCTP packet structure is designed to offer the benefits of connection-oriented data flow (sequential) with the variable packet size and the use of Internet protocol (IP) addressing.

Figure 14.6 shows that the SCTP packet structure includes a common header format with chunks of data. The header includes the source and destination port numbers that are associated with the specific IP addresses to uniquely identify the packet for a specific communication application. The verification tag uniquely identifies (validates) the sender of the packet. Each packet has a checksum to ensure all the data has been reliably sent. The data is sent in chunks to allow the near real-time streaming (continuous one-way delivery) of information.

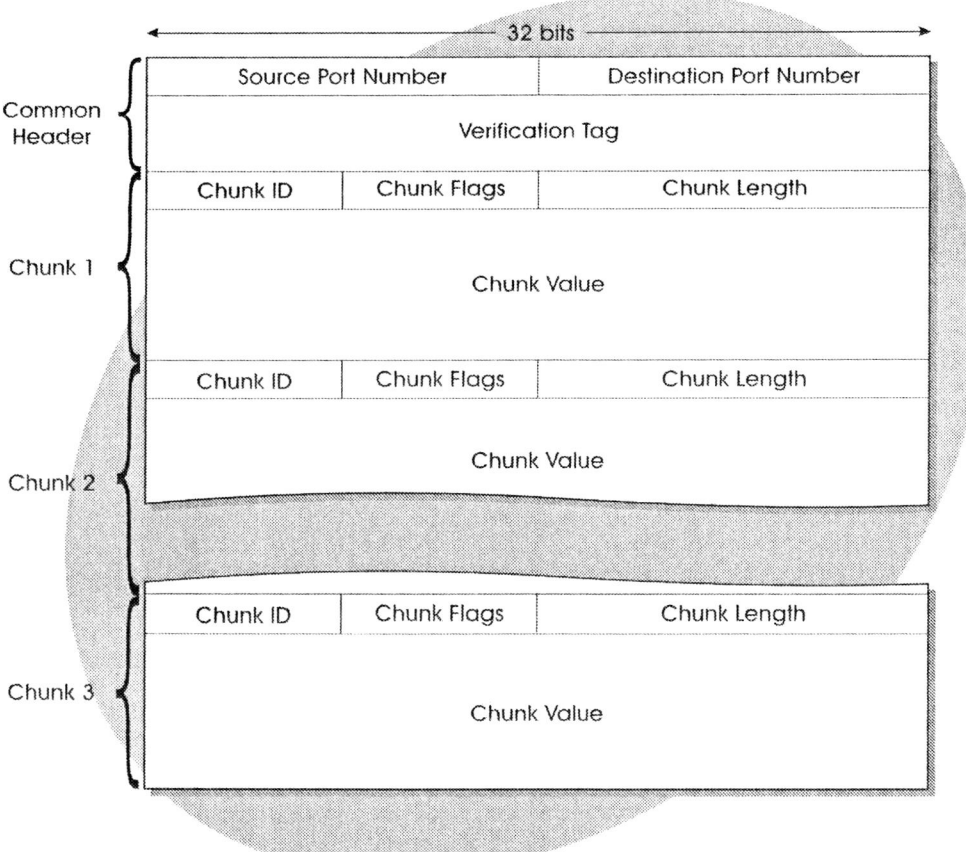

Figure 14.6, SCTP Packet Structure

MTP3 User Adaptation Layer (M3UA)

M3UA MTP3-User Adaptation Layer is a protocol for supporting the transport of any SS7 MTP3-User signaling (e.g., ISUP and SCCP messages) over Internet Protocol (IP) using the services of the Stream Control Transmission Protocol (SCTP). This protocol would be used between a Signaling Gateway (SG) and a Media Gateway Controller (MGC) or IP-resident Database. The M3UA layer would be used to adapt protocols for Application Servers (ASs) that may be running a set of Application Server Processes (ASPs) or it may be used to provide interfaces for routing keys (OPC, DPC, and SLS).

Figure 14.7 shows how the M3UA protocol adaptation layer can be used to simulate the functions of the MTP3 layer of the SS7 system. This diagram shows that the M3UA layer is used to adapt SS7 signaling from the SCCP layer into a transport layer that can be routed using IP.

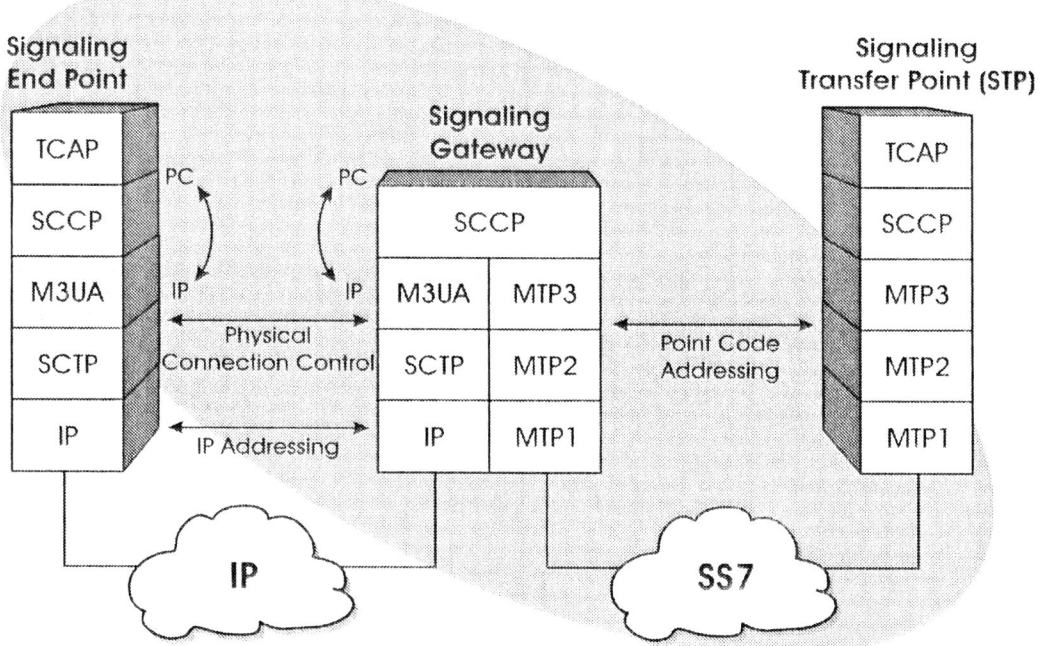

Figure 14.7. M3UA Protocol Adaptation Layer

MTP2 User Adaptation Layer (M2UA)

The MTP2 User Adaptation Layer (M2UA) is a protocol for supporting the transport of SS7 MTP2-User signaling (e.g., ISUP call setup messages) over Internet Protocol (IP) using Simple Control Transmission Protocol (SCTP). This M2UA would be used in systems that only require MTP3 communication. The MTP2-User Adaptation Layer protocol would be used between a Signaling Gateway (SG) and a Media Gateway Controller (MGC) or IP-resident Database.

Figure 14.8 shows how the M2UA protocol adaptation layer can be used to simulate the functions of the MTP3 layer of the SS7 system. This diagram shows that the M3UA layer is used to adapt SS7 signaling from the SCCP layer into a transport layer that can be routed using IP.

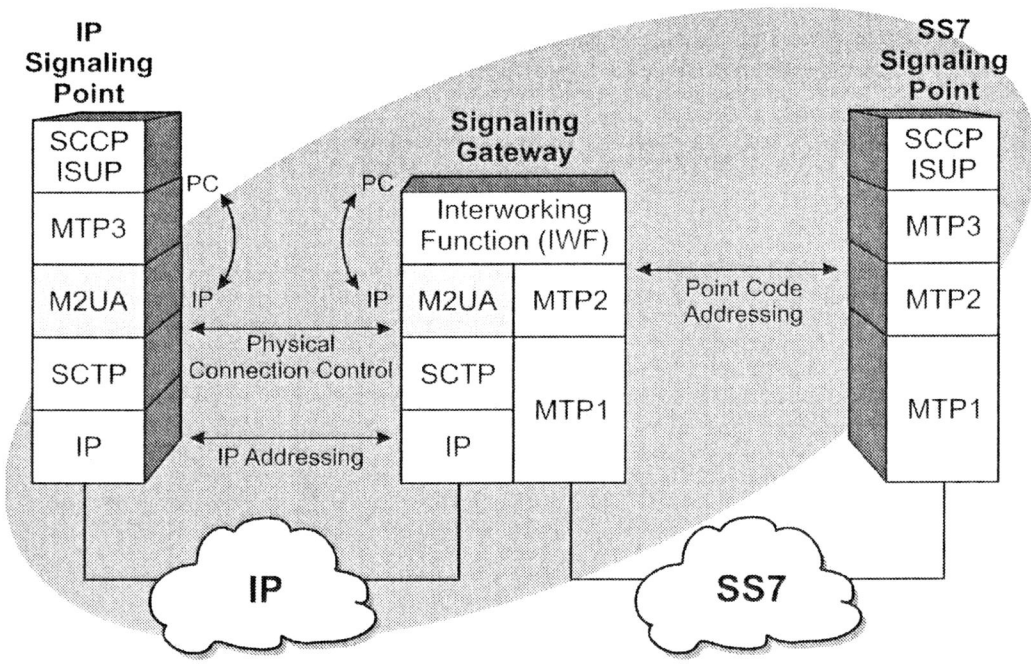

Figure 14.8, M2UA Protocol Adaptation Layer

MTP2 Adaptation Layer (M2PA)

MTP2 Adaptation Layer (M2PA) is a protocol for supporting the transport of any network MTP3 signaling over Internet Protocol (IP) using the services of the Stream Control Transmission Protocol (SCTP). MTP2 allows for seamless operation of MTP3 messages regardless if the network is SS7 or IP. M2PA also allows for the transmission of network management messages.

Figure 14.9 shows how the M2PA protocol adaptation layer provides full MTP3 message handing protocol between SS7 nodes connected by IP based networks.

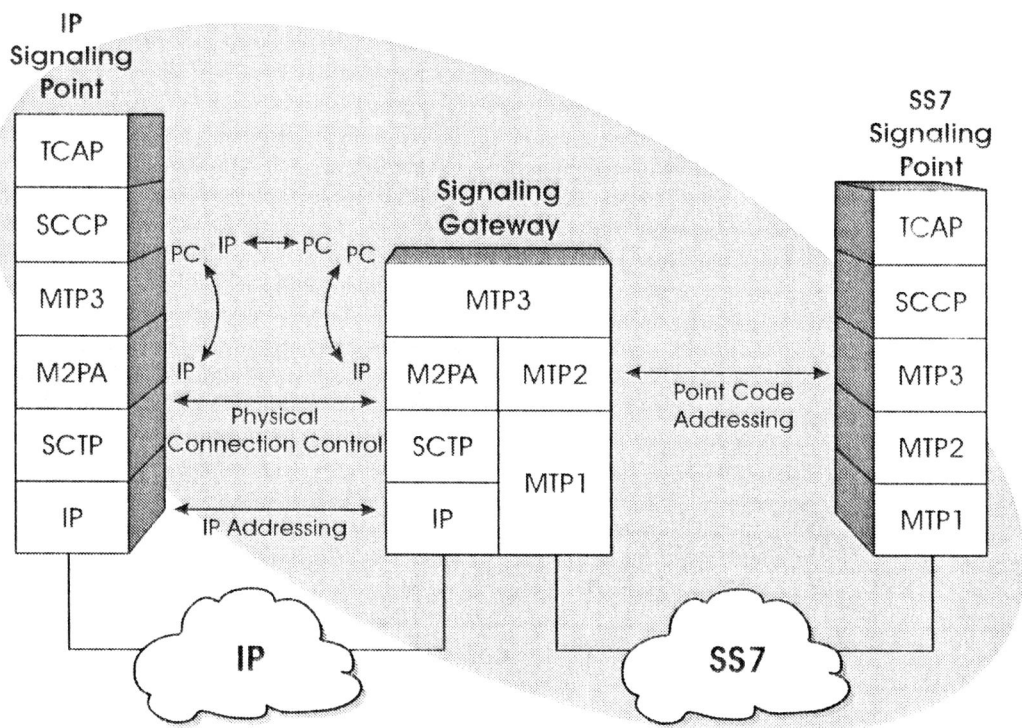

Figure 14.9. M2PA Protocol Adaptation Layer

CCP User Adaptation Layer (SUA)

SCCP User Adaptation layer is a protocol for the transport of any SS7 SCCP user signaling (e.g. TCAP or MAP messages) over IP Between two signaling endpoints. The use of an SUA protocol eliminates the use of the MTP protocol portion in a signaling system.

Figure 14.10 shows the basic operation of the SCCP User Adaptation (SUA) protocol. This diagram shows that the SUA protocol allows the upper layer signaling applications to directly communicate with the IP based SCTP signaling transport protocol. This diagram shows that the SCTP protocol maintains the connection control through the network while the SUA protocol maps the communication channel from the upper layer application to the transport channel over the IP network.

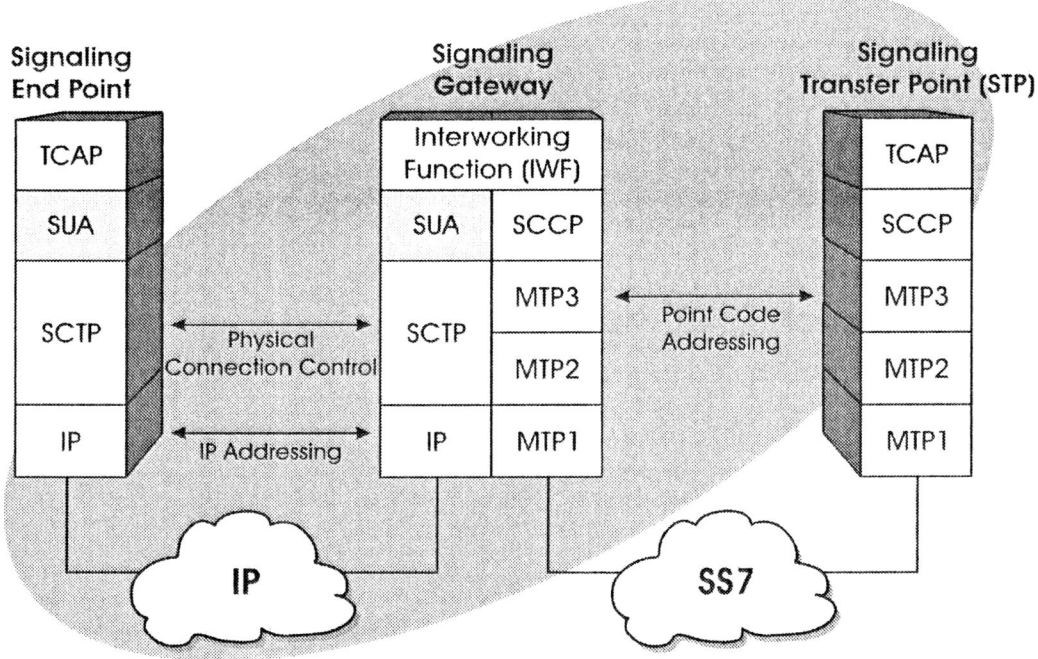

Figure 14.10, SCCP User Adaptation (SUA) Layer

ISDN User Adaptation Layer (IUA)

ISDN User Adaptation layer is used to transport ISDN user signaling (Q.931) over IP between two signaling endpoints. The use of an IUA protocol eliminates the use of the MTP protocol portion in a signaling system.

Figure 14.11 shows the basic operation of the ISDN User Adaptation (IUA) protocol. This diagram shows that the IUA protocol allows the upper layer signaling applications to directly communicate with the IP based SCTP signaling transport protocol. This diagram shows that the SCTP protocol maintains the connection control through the network while the SUA protocol maps the communication channel from the upper layer application to the transport channel over the IP network.

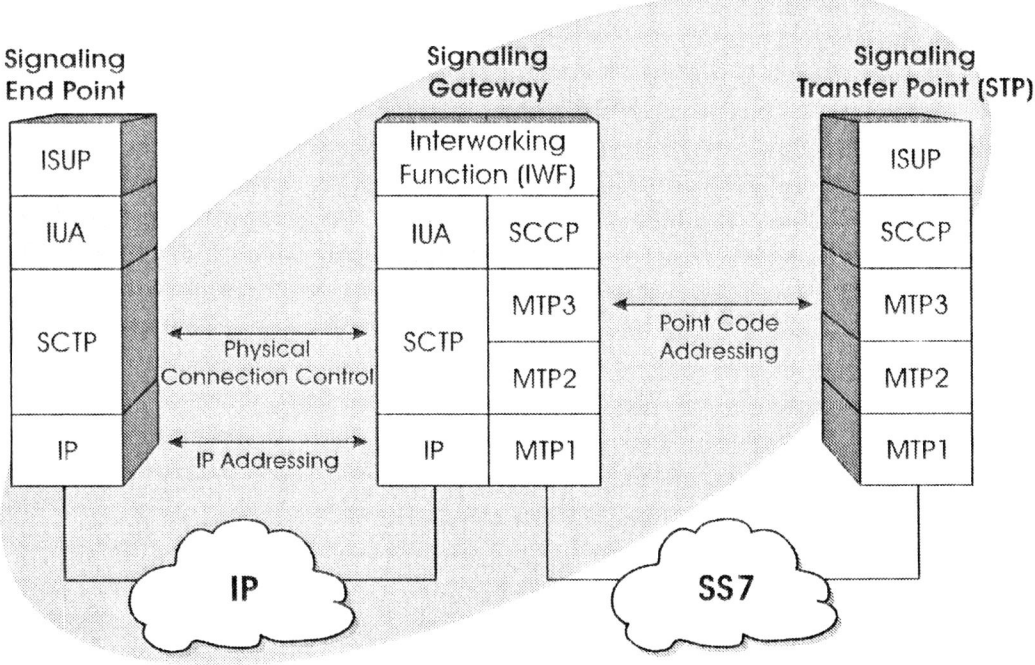

Figure 14.11. ISDN User Adaptation (IUA) Layer

Internet Protocol (IP) Telephony Systems

Internet telephony provides voice services through the use of the public Internet. These Internet telephony systems initiate, process, and receive voice (and sometimes multimedia) communications using the embedded protocol of the Internet. These systems may use the public Internet, private data networks (e.g. LAN based), private IP-based networks or hybrids of public and private systems.

IP telephone systems require the use of signaling messages to setup, maintain, and disconnect communication systems. The signaling technologies used for IP telephone systems may be interfaced with the SS7 signaling system to allow the interconnection of IP telephone systems to public telephone networks. The common signaling protocols for IP telephony systems include media gateway control protocol (MGCP), MEGACO, session initiation protocol (SIP), and H.323.

Media Gateway Control Protocol (MGCP)

MGCP is a control protocol that uses a controller to setup, manage, and terminate multimedia communication sessions in a centralized communications system through media gateways. In essence, a system that uses MGCP has centralized control. This centralized control structure differs from other multimedia control protocol systems (such as H.323) that allow distributed network control as end points in the network may directly setup and control a communication session. MGCP is specified in RFC 2705 and it was first drafted in 1998. MGCP forms the basis of the PacketCable NCS protocol.

MEGACO is a control protocol that is similar to MGCP and some industry experts believe that MEGACO will eventually replace MGCP. MCGP can use either text format or binary format messages to setup, manage, and terminate multimedia communication sessions. MEGACO offers the potential for large distances between the media gateway controller (MGC) and media gateway (MG).

Session Initiation Protocol (SIP)

SIP is an application layer protocol that uses text format messages to setup, manage, and terminate multimedia communication sessions. SIP is a simplified system as compared to the ITU H.323 packet multimedia system. SIP is defined in RFC 2543.

The SIP system is primarily composed of clients (equivalent of a media gateway) and servers (equivalent of media gateway controller or gatekeepers). SIP protocol can use session description protocol (SDP) to control media streams through a network. SDP is a text-based protocol that is used throughout to provide high-level definitions of connections and media streams. The SDP protocol is used with session Initiation protocol (SIP). The SDP protocol is used in the PacketCable system. SDP is defined in RFC 2327.

Figure 14.12 shows the SS7 that is used to initiate a call into an SIP system. This diagram shows that a caller that is connected to a SIP network (SIP client) initiates a call using an invite command that contains a destination SIP Uniform Resource Locator (URL). This identifier is sent to the proxy server that determines and maps the URL to the actual destination number that will be sent in the initial address message (IAM). The proxy server informs the SIP client that the call routing is in progress (it is trying to connect). The Invite command is then forwarded to the network gateway (NGW). The NGW creates an IAM that contains the destination phone number. The PSTN switch sends back an ACM to the NGW. The NGW informs the proxy server that the call is progressing and the proxy server forwards this session progress message to the SIP client. This allows an audio path to be connected between the PSTN and the SIP client. When the destination telephone user answers, the PSTN sends and answer message to the NGW. The NGW translates this command and sends a message updating the session to indicate the call has been answered. This is forwarded to the SIP client. When the SIP client acknowledges the message, the NGW can connect a second media path from the SIP client to the PSTN switch.

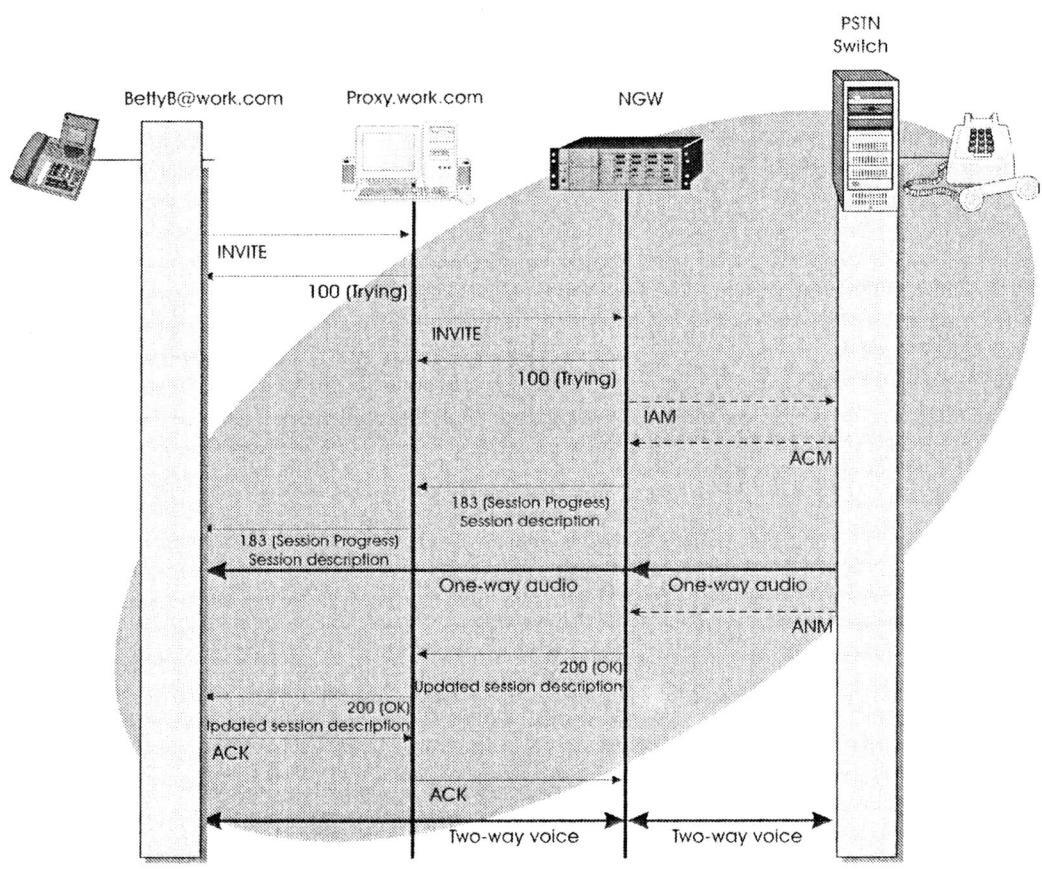

Figure 14.12. SS7 Signaling into an SIP System

H.323

H.323 is an umbrella recommendation from the International Telecommunications Union (ITU) that sets standards for multimedia communications over Local Area Networks (LANs) that may not provide a guaranteed Quality of Service (QoS). H.323 specifies techniques for compressing and transmitting real-time voice, video, and data between a pair of video-conferencing workstations. It also describes signaling protocols for managing audio and video streams, as well as procedures for breaking data into packets and synchronizing transmissions across communications channels. The H.225 protocol is used for call signaling for the set up of connections between H.323 endpoints (terminals and gateways), over which the real-time data can be transported. Call signaling involves the exchange of H.225 protocol messages over a reliable call-signaling channel. For example, H.225 protocol messages are carried over TCP in an IP based H.323 network.

After a communication session is established, H.245 protocol can be used to send control signaling between communicating H.323 endpoints. The H.245 control messages are carried over H.245 control channels. The H.245 control channel is the logical channel 0 and is permanently open, unlike the media channels. The messages carried include messages to exchange capabilities of terminals and to open and close logical channels.

Supplementary Services for H.323, namely Call Transfer and Call Diversion, have been defined by the H.450 series of specifications. H.450.1 defines the signaling protocol between H.323 endpoints for the control of supplementary services. H.450.2 defines Call Transfer and H.450.3 Call Diversion. Call Transfer allows a call established between endpoint A and endpoint B to be transformed into a new call between endpoint B and a third endpoint, endpoint C. Call Diversion provides the supplementary services Call Forwarding Unconditional, Call Forwarding Busy, Call Forwarding No Reply and Call Deflection. The "hooks" for Supplementary Services are specified in H.323 Version 2.

The H.323 system can interconnect standard telephones and data communication devices (multimedia computers) through the use of gateways and gatekeepers. Gateways convert the audio and multimedia information into formats that can be transmitted through the packet data network. Gatekeepers coordinate, authorize, and bill (if billing is required) access through the gateways. When calls are initiated through the H.323 network, the gateway requests access from the gatekeeper. The gatekeeper reviews its database to determine if the request is authorized and may perform translation of dialed digits to a data (IP) address. The destination gatekeeper is then contacted and if it authorizes service, its associated gateway will be setup to translate the call to the communication device (e.g. telephone).

Figure 14.13 shows a sample of how call signaling may progress when a PSTN telephone user initiates a call to a terminal in an H.323 system. In this example, the caller dials the H.323 telephone number. This produces an initial address message (IAM) to the H.323 gateway. This creates an H.323 access request message (ARQ) to the gatekeeper to lookup the destination IP address. The gatekeeper responds with an Admission Confirm (ACF) message that provides the gateway with the destination address of the H.323 terminal, (IP Telephone). In this example, the H.323 gateway sends a setup messages (fast setup that includes the initial call parameters) and the H.323 terminal responds with a call proceeding message. This allows the H.323 gateway to create an SS7 CPG message to the originating PSTN switch. The H.323 terminal sends an access request (ARQ) message to the gatekeeper. The gatekeeper responds with an access confirmation (ACF) message that allows the call to proceed. This results in the H.323 terminal sending an alerting message that indicates that the terminal is ringing. This permits the H.323 gateway to send an SS7 ACM message to the PSTN switch. The media path from the H.323 terminal is then connected to the audio channel of the PSTN switch (packet voice is converted to PCM circuit voice in the H.323 gateway). When the user at the H.323 terminal answers the call, the terminal sends a connect message (fast setup that includes parameters) to the H.323 gateway. This alerts the PSTN switch with an SS7 answer message (ANM). This allows a second media path to be connected between the PSTN switch and the H.323 terminal.

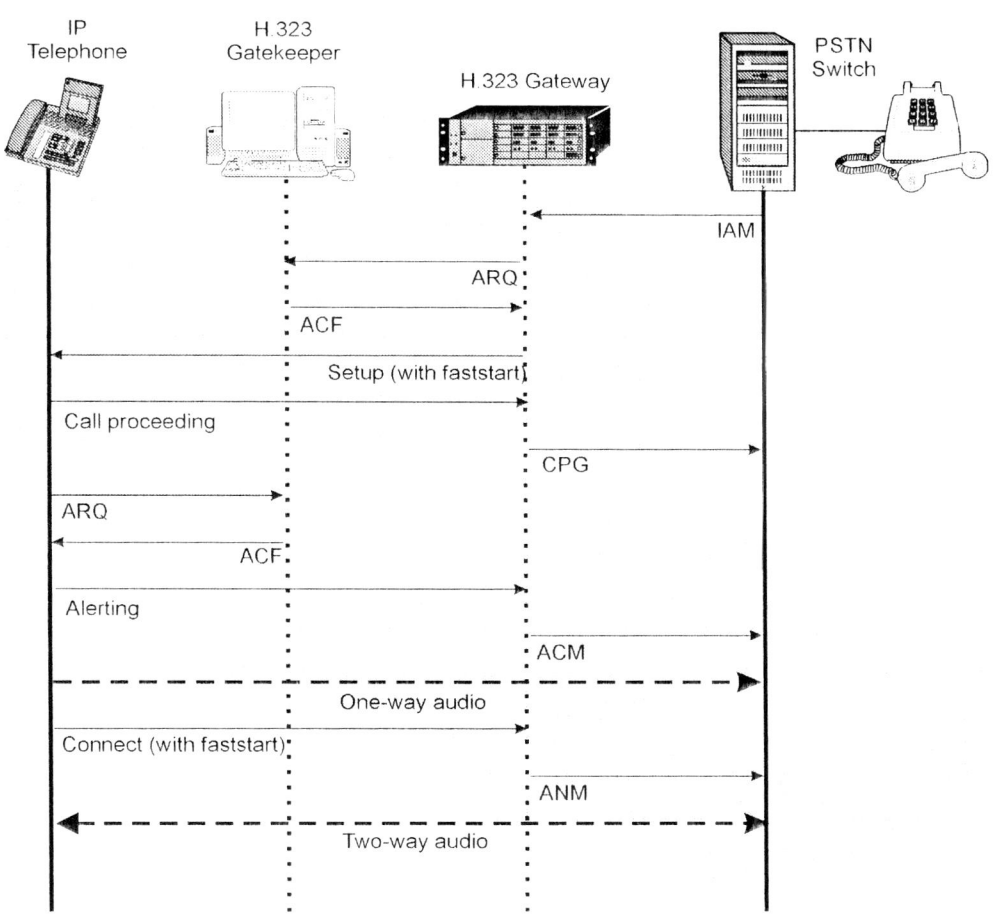

Figure 14.13. SS7 Signaling into an H.323 System

Chapter 15

International SS7

There are different versions of SS7 systems that are used throughout the world. Some of the variations of SS7 systems include different point code addressing, circuit identification code (CIC) formats, transcoding control requirements, and international gateway support.

SS7 Addressing

SS7 addressing involves the use of destination point code (DPC) and origination point code (OPC). The use of point codes identifies specific equipments and possibly functional assemblies within these equipments. The number of bits assigned to point codes by ANSI is 24 and by ITU is 14. The assignment of these point codes to specific operators and system also varies. ANSI divides their point codes into 8 bit groups of member, cluster, and network groups. This allows each operator that uses ANSI to have at least 65,536 unique point code addresses.

Figure 15.1 shows the basic differences between the SS7 MSUs packet structures as defined by ANSI and the ITU. This diagram shows that the ITU version uses a destination point code (DPC) and origination point code (OPC) that are 14 bits long. The ANSI MSU version uses a DPC and OPC codes that are 24 bits. The ANSI DPC and OPC codes are also divided into hierarchical structure that has member, cluster, and network groups of 8 bits each.

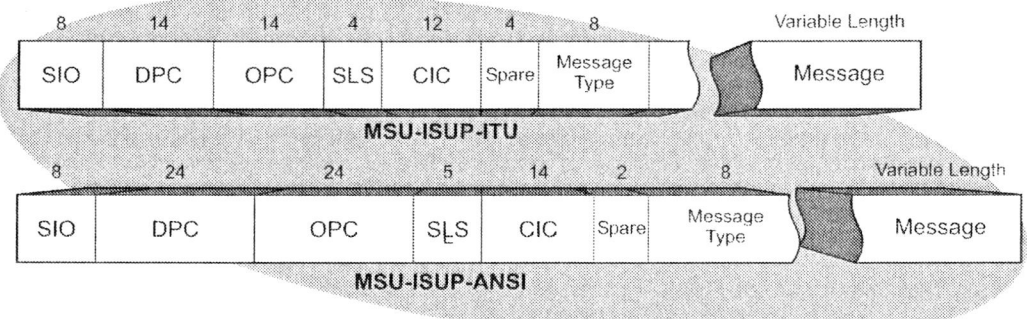

Figure 15.1. SS7 Addressing Differences between ANSI and ITU

Circuit Identification Codes (CICs)

The circuit identification code (CIC) is an information code that identifies a circuit between a pair of SS7 exchanges for which signaling is being performed (14 bits in the ANSI version and 12 bits in the ITU version ISDN User Part.)

Figure 15.2 shows the basic differences between the SS7 MSUs used for ISDN user part (ISUP) messages. This diagram shows that the ITU version uses a circuit identifier code (CIC) that is 12 bits long and the ANSI MSU for ISUP uses a CIC code that has 14 bits.

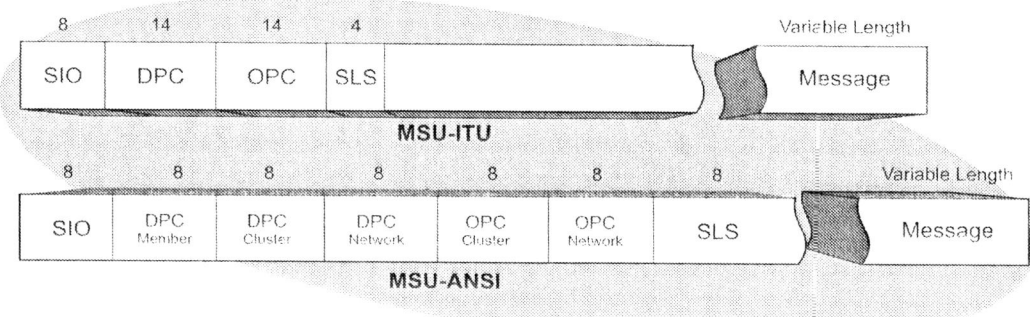

Figure 15.2. SS7 Circuit Identification Code Differences between ANSI and ITU

Transcoding

Transcoding is the conversion of digital signals from one coding format to another. A common example of a transcoding application is the conversion of μ-Law encoded PCM to A-Law encoded PCM signals. This is necessary to interconnect systems that use different types of PCM coding (such as the Americas and European systems).

Figure 15.3 shows that transcoding process that is used to convert A-Law to μ-Law PCM signals. This diagram shows that a telephone using A-law PCM speech coding in a European System is communicating with a telephone in North America that is using μ-law PCM speech coding. This diagram shows that the transcoding system must identifies the type of PCM audio used by each system and the location of the transcoding gateway function. The PCM transcoder converts the A-Law PCM signal to μ-Law PCM.

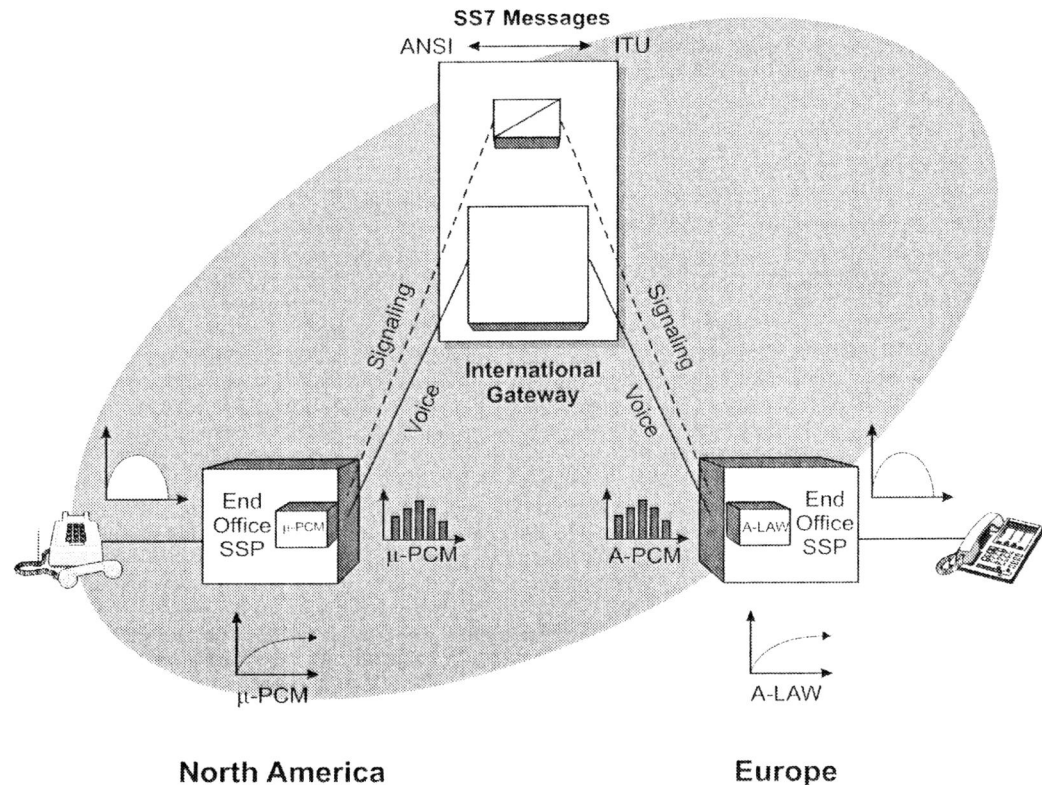

Figure 15.3, Transcoding A-Law and μ-Law

International Gateway Facilities (IGF)

International gateway facilities (IGF) are systems or equipment that provide access between telephone systems in different countries. International gateways may convert SS7 and other signaling formats between different signaling formats. These include ANSI standards, ITU standards, national variants of SS7 signaling standards, MF signaling, and R2 signaling. International gateways may also provide for transcoding services between μ-Law PCM and A-Law PCM speech coding.

Figure 15.4 shows how two national SS7 systems interconnect using an international gateway between an ANSI based end office SSP in North America and an ITU based end office SSP in Asia. This example shows that an ANSI based SS7 system require address translation and circuit identifier code format changes as the messages are passed between the systems. The ANSI 24 bit destination point code (DPC) and origination point code (OPC) addressing must be translated to 14 bit DPC and OPC codes for the ITU system. It also shows that the 14 bit ANSI CIC code used in ISUP messages must be translated to 12 bit CIC codes used by the ITU system.

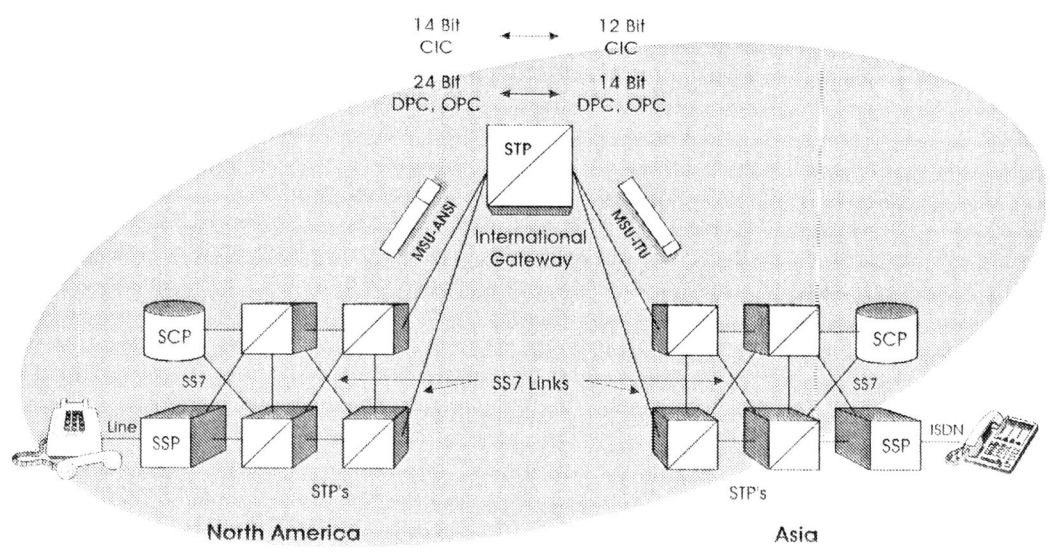

Figure 15.4. International Gateways

Glossary

800 Service-A communication services where the receiving party pays for incoming calls. 800 services is often called "Toll Free" In the Americas and "Freephone" in Europe and the dialing digits are 0800. In the United States, additional toll free numbers include 888, 877, and 866.

A & B Signaling-A process of transferring signaling information on a digital communication line that involves the stealing of bits from the digital transmission line (called the A & B bits). This is a type of in-band signaling as it steals some of the channel bandwidth for use as signaling messages. When it is used in T1 transmission line, 1 bit of user data from each of the 24 sub-channels in every sixth frame is stolen (discarded) and replaced with signaling control information. The use of A+B signaling reduces the T1 bandwidth from 1.544Mbps to 1.536 Mbps.

AAL-See: ATM Adaptation Layer

AAL1-See: ATM Adaptation Layer 1

Access Link (A-Link)-A communications line (link) that provides access from a service switching point (SSP) and signaling control point (SCP) to signaling transfer points (STPs) in an SS7 network.

Access Tandem (AT)-A high level switching system that interconnects low level (local exchange) switching systems. An access tandem can also provide access for nonconforming end offices such as for equal access to other long distance service providers.

ACD-See: Automatic Call Distribution

ACM-See: Address Complete Message

Active Signaling Link-A signaling link that is being used in the Signaling System 7 system that has completed the initial alignment procedures and is transporting or ready to transport signaling messages.

Adaptive Differential Pulse Code Modulation (ADPCM)-A process of converting analog voice signals into encoded digital signals through the use of predictive codes that are created by analyzing the previous digital audio signals. ADPCM is derived from the original pulse code modulation (PCM) system that commonly represents an analog signal as 64 kbps (called a DS0). ADPCM systems commonly provide digital signals at 32 kbps and 16 kbps.

ADC-See: Analog to Digital Converter

Address Complete Message (ACM)-An ISDN User Part trunk signaling message by which a call's destination SSP acknowledges an initial address message.

Address Translation Gateway-A gateway that can convert the address format (field and/or physical layer structure) from one network to another.

Adjunct Processor (AP)-Adjunct Processor (AP) is a decentralized form of signaling control point (SCP) databases that are used by switching systems to reduce the requirement of signaling service point (SSP) central office switches from connecting to SCPs.

ADPCM-See: Adaptive Differential Pulse Code Modulation

ADSI-See: Analog Display Services Interface

ADSL-See: Asymmetric Digital Subscriber Line

Advanced Intelligent Network (AIN)-Advanced intelligent networks (AIN's) are telecommunications networks that are capable of providing advanced services through the use of distributed databases that provide additional information to call processing and routing requests.

In the mid 1980's, Bellcore (now Telcordia) developed a set of software development tools to allow companies to develop advanced services for the telephone network. The advanced intelligent network (AIN) is a combination of the SS7 signaling network, interactive database nodes, and development tools that allow for the processing of signaling messages to provided for advanced telecommunications services.

The AIN system uses a service creation environment (SCE) to created advanced applications. The SCE is a development tool kit that allows the creation of services for an AIN that is used as part of the SS7 network. A service management system (SMS) is the interface between applications and the SS7 telephone network. The SMS is a computer system that administers service between service developers and signal control point databases in the SS7 network. The SMS system supports the development of intelligent database services. The system contains routing instructions and other call processing information.

To enable SCPs to become more interactive, intelligent peripherals (IPs) may be connected to them. IPs are a type of hardware device that can be programmed to perform a intelligent network processing for the SCP database. IPs perform processing services such as interactive voice response (IVR), selected digit capture, feature selection, and account management for prepaid services.

To help reduce the processing requirements of SCP databases in the SS7 network, adjunct processors (APs) may be used. APs provide some of the database processing services to local switching systems (SSPs).

AERM-See: Alignment Error Rate Monitor

AIN-See: Advanced Intelligent Network

A-law Encoding-A digital signal companding process that is used for encoding/decoding signals in pulse-code-modulated (PCM) systems. This companding process increases the dynamic range of a binary signal by assigning different weighted values to each bit of information than is defined by the binary system. The A-law encoding system is an international standard. A different companding version is used in the Americas is uLaw.

Algorithm-A set of well-defined steps or rules that allows for the solution of a problem or processing of information. Commonly the name for a portion of a software program.

Alignment Error Rate Monitor (AERM)-An error rate monitor that is used in the SS7 system that is used to estimate the error rates associated with a signaling link during the initial alignment (link activation) period.

A-Link-See: Access Link

Alternative Routing Of Signaling-The routing of signaling messages (such as in a Signaling System 7 system) through alternative paths as a result in the failure of a primary routing path.

AMA-See: Automatic Message Accounting

American National Standards Institute (ANSI)-The US organization that sets the rules and procedures for, and also authorizes specific standards setting organizations. ATIS and EIA/TIA are two ANSI authorized standards setting organizations in the US in the subject area of telecommunications.

AMIS-See: Audio Messaging Interchange Specification

Analog Display Services Interface (ADSI)-A Bellcore/Telcordia industry standard that defines the flow of information between a network element (e.g. network server, switch, voice mail system) and a customer's telephone, PC or other terminal device that contains a screen display.

Analog to Digital Converter (ADC)-A signal converter that changes a continuously varying signal (analog) into a digital value. A typical conversion process includes an initial filtering process to remove extremely high and low frequencies that could confuse the digital converter. A periodic sampling section that at fixed intervals locks in the instantaneous analog signal voltage, and a converter that changes the sampled voltage into its equivalent digital number or pulses.

ANI-See: Automatic Number Identification

ANM-See: Answer Message

ANSI-See: American National Standards Institute

Answer Message (ANM)-An ISDN User Part trunk signaling message that indicates the called party has answered the call.

AP-See: Adjunct Processor

Application Layer-The application layer protocol coordinates the information interface between the communication device and the end user. The application layer receives data from the underlying protocols and processes this information into a form required or requested by the user or endpoint device. The application layer usually requests or responds to requests for a communication session. The location of the application layer is at the top of protocol stacks. The application layer is layer 7 in the open system interconnection (OSI) protocol layer model.

Application Service Element (ASE)-A software program or portion of a communication protocol that is part of an application layer of a protocol stack. Several ASEs may be combined to form a complete application protocol.

Application Service Provider (ASP)-A company that provides an end user with an information service. An ASP owns are leases computer hardware and software system that allows one or more users to access information services on or through that computer systems.

ASE-See: Application Service Element

ASP-See: Application Service Provider

Asymmetric Digital Subscriber Line (ADSL)-A communication system that transfers both analog and digital information on a copper wire pair. The analog information can be a standard POTS or ISDN signal. The maximum downstream digital transmission rate (data rate to the end user) can vary from 1.5 Mbps to 9 Mbps downstream and the maximum upstream digital transmission rate (from the customer to the network) varies from 16 kbps to approximately 800 kbps. The data transmission rate varies depending on distance, line distortion and settings from the ADSL service provider.

Asymmetrical Private Virtual Circuit (Asymmetrical PVC)-A virtual circuit that permits uneven (asymmetrical) data transmission rates for each direction of transmission.

Asymmetrical PVC-See: Asymmetrical Private Virtual Circuit

Asynchronous Transfer Mode (ATM)-A packet data and switching technique that transfers information by using fixed length 53 byte cells. The ATM system uses high-speed transmission (155 Mbps) and is a connection-based system. When an ATM circuit is established, a patch through multiple switches is setup and remains in place until the connection is completed. ATM service was developed to allow one communication medium (high speed packet data) to provide for voice, data and video service.

As of the 1990's, ATM has become a standard for high-speed digital backbone networks. ATM networks are widely used by large telecommunications service providers to interconnect their network parts (e.g. DSLAMs and Routers). ATM aggregators operate networks that consolidate data traffic from multiple feeders (such as DSL lines and ISP links) to transport different types of media (voice, data and video).

AT-See: Access Tandem

ATM-See: Asynchronous Transfer Mode

ATM Adaptation Layer (AAL)-A set of standard protocols that translate user traffic into a size and format that can be contained in the small payload of an ATM cell. User traffic is returned to its original form at the destination. This process is called segmentation and reassembly. All AAL functions occur at the ATM end-station rather than at the switch. These protocol layers are designed to allow for constant bit rate (CBR), variable bit rate (VBR), unspecified bit rate (UBR), and other types of services.

ATM Adaptation Layer 1 (AAL1)-The layer within the ATM protocol that converts the 53 byte packets from the network into the form used by the customer for constant bit rate (CBR) services.

Attendant Call Waiting Indication-The indication light or message on the attendant console that indicates that one or several calls are in queue to be answered. The indication may change (e.g. flash or ring) when additional thresholds (e.g. maximum number of waiting calls) are reached.

AUC- Authentication Center

Audio Messaging Interchange Specification (AMIS)-An industry specification that defines how messages are exchanged on a network between voice mail systems.

Authentication Center (AUC)-A part of a network that manages the encryption keys that validate the identity of customers and enable voice privacy services. A single authentication center may process validation requests using different keys, random numbers and encryption algorithms.

Automatic Call Distribution (ACD)-ACD is a system that automatically distributes incoming telephone to specific telephone sets or stations calls based on the characteristics of the call. These characteristics can include an incoming phone number or options selected by a caller using an interactive voice response (IVR) system. ACD is the process of management and control of incoming calls so that the calls are distributed evenly to attendant positions. Calls are served in the approximate order of their arrival and are routed to service positions as positions become available for handling calls.

Automatic Callback-A CLASS service feature that allows a caller to complete a call to a busy station by dialing an activation code (usually a single digit) and hanging up. The system automatically rings both parties when the lines are available.

Automatic Message Accounting (AMA)-An automatic system for recording data describing the origination time of day, dialed number and time duration of a call for purposes of billing. The earliest systems used punched paper tape, later replaced by magnetic computer tape and then later magnetic computer disk. AMA is a term mostly used in the public network, and similar terms, some used in private, PBX, or inter-carrier systems are Call Detail Recording (CDR), Station Detail Message Recording (SMDR), and Automatic (calling) Number Identification (ACNI or ANI).

Automatic Number Identification (ANI)-(1-general) the capability in a telephone system to identify the originating telephone number, including an extension number in a PBX or Centrex system, usually for billing purposes but also for use by automatic call back and other features. (2-number display) The number of the line or station thus identified. [3-auto number identifier) Want to know what number you're calling from? Call 1-800-346-0152 (does not work on all phones). See also related terms AIOD and AMA and CDR.

Backward Indicator Bit (BIB)-A bit in an SS7 signaling message, by its status change at the remote end, requests retransmission because of message received out of sequence.

Backward Sequence Number (BSN)-A field in an SS7 signal unit sent that contains the forward sequence number of a correctly received signal unit being acknowledged.

Bandwidth-A term that is applicable to signals that occupy a portion of a frequency spectrum, particularly a radio system. Analysis or measurement of the signals or signal waveforms of such a system will show that most (or substantially all) of the power contained in that signal can be found in a designated portion of the frequency spectrum. The difference between the highest and lowest frequency describing that portion of the spectrum is the bandwidth of the signal. Bandwidth is measured in units of hertz or cycles per second.

Bandwidth On Demand (BoD)-A system that allows different data transmission rates based on requests from the customer, their application (e.g. voice or video), and the data transmission capability of the system.

Baud-Name for the unit of data symbols per second. This name, abbreviated Bd, is taken from the name of the 19th century French teletypewriter machine innovator Emiel Baudot. For a method of modulation or encoding in which there is a choice of only two symbol values per symbol interval, or one bit per symbol (such as two-level pulse voltages) the baud rate is equal to the bit rate (bits per second). For a method of modulation or encoding in which there are more than two symbol values per symbol interval (and thus 2 or more bits per symbol) the bit rate is higher than the baud rate. For example, QPSK phase modulation and 2B4Q pulse coding both have 4 symbol values per symbol interval and thus the bit rate (bits per second) is twice the symbol (baud) rate. (Please do not make the error of writing "baud per second.")

BIB-See: Backward Indicator Bit

B-ISDN-See: Broadband Integrated Services Digital Network

B-ISUP-See: Broadband Integrated Services Digital Network User Part

Bit-In digital signals, a bit is a single unit of data. Generally, a bit has a value of either one or zero corresponding to on or off for an optical or electrical signal. Bits are typically clustered into groups of eight, called bytes or octads. Different coding schemes, such as ASCII, have been developed to assign meaning to the bytes.

B-Link-See: Bridge Link

BoD-See: Bandwidth On Demand

Bong-A unique tone that telephone carriers and to an audio signal while a call is in progress to alert the user that another action on the users part is required (such as to dial an access code number for a calling card).

Bridge Link (B-Link)-A "B" (bridge) link connects a signal transfer point (STP) to another STP in an SS7 network. Typically, a quad of "B" links interconnect peer (or primary) STPs (e.g., the STPs from one network to the STPs of another network). The distinction between a "B" link and a "D" link is rather arbitrary. For this reason, such links may be referred to as "B/D" links.

Broadband Integrated Services Digital Network (B-ISDN)-Usually refers to the portions of a digital network operating at data transfer rates in excess of 1.544 or 2.048 Mbps. The B-ISDN network often uses ATM to enables transport and switching of voice, data, image, and video over the same network equipment.

Broadband Integrated Services Digital Network User Part (B-ISUP)-An SS7 protocol that defines the signaling messages that are used to control ATM broadband connections and services.

BSN-See: Backward Sequence Number

Busy Hour-The hour in a day when the total usage of the network, trunk connection, or the switching system is greater than at any other hour during the day. Telephone systems and networks are typically designed to meet a specific quality level that can be provided during the busy hour (e.g. maximum number of blocked calls).

CABS-See: Carrier Access Billing System

Call Center-A call center is a place where calls are answered and originated, typically between a company and a customer. Call centers assist customers with requests for new service activation and help with product features and services. A call center usually has many stations for call center agents that communicate with customers. When call agents assist customers, they are typically called customer service representatives (CSRs).

Call centers use telephone systems that usually include sophisticated automatic call distribution (ACD) systems and computer telephone integration (CTI) systems. ACD systems route the incoming calls to the correct (qualified) customer service representative (CSR). CTI systems link the telephone calls to the accounting databases to allow the CSR to see the account history (usually producing a "screen-pop" of information).

Call Detail Record (CDR)-A data record that holds information related to a telephone call. This information usually contains the origination and destination address of the call, time of day the call was connected, added toll charges through other networks, and duration of the call.

Call Forward Busy-The process of forwarding a call to another extension or telephone number when the selected extension or telephone number is in use (busy).

Call Forwarding-A call processing feature allows a user to have telephones calls automatically redirected to another telephone number or device (such as a voice mail system). There can be conditional or unconditional reasons for call forwarding. If the user selects that all calls are forwarded to another telephone device (such as a telephone number or voice mailbox), this is unconditional. Conditional reasons for call forwarding include if the user is busy, does not answer or is not reachable (such as when a mobile phone is out of service area).

Call Pickup-A telephone call processing (switch control) feature that enables a telephone user to answer a telephone from another telephone station. When a call is received, a key sequence is entered from specific groups or any telephone (dependent on how the system is setup) and the call is redirected to the extension or line that has picked up (entered the code) the line.

Call Processing-Steps that occur during the duration of a call. These steps are typically associated with the routing and control of the call. When used as part of a billing system, call process involves gathering messages from various sources (event records), reformatting and edit the messages, calculate a charge for the message, assign a customer to the message, and getting the message ready for billing.

Call Rating-A process of assigning a value or cost to a telephone call or communication session.

Call Trace-Call trace allows a subscriber to initiate a call trace request message that allows the dialed digits of a caller to be stored for investigation. The activation of the call trace service alerts the telephone service operator to "tag" the originator's number to allow authorities to investigate the originator of the unwanted or unauthorized call. Some of the call trace activation codes include *57 on a touchtone phone or the dialing of 1157 on a rotary (pulse) phone. If the call trace of the last call was completed successfully, an announcement should be heard. The service operator will usually release the call trace information to law enforcement agencies and a signed authorization from the subscriber may be required.

Call Waiting (CW)-A telephone call processing feature that notifies a telephone user that a another incoming call is waiting to be answered. This is typically provided by a brief tone that is not heard by the other callers. Some advanced telephones (such as digital mobile telephones) are capable of displaying the incoming phone number of the waiting call.

After the service provides the subscriber with the notification of an incoming call while the subscriber's call, controlling subscriber can either answer or ignore the incoming call. If the controlling subscriber answers the second call, it may alternate between the two calls.

Called Line Identification-The identification information carried by an SS7 packet that provides the destination receiver to identify the source of the call.

Calling Card-An identifying number or code unique to the individual, that is issued to the individual by a common carrier and enables the individual to be charged by means of a phone bill for charges incurred independent of where the call originates.

Calling Line Identification (CLI)-A service which displays the calling number prior to answering the call that allows telephone customer to determine if they want to answer the call. The calling number may be used by the telephone device to look-up a name in memory (e.g. mom) and display the name along with the phone number.

Calling Number Display Blocking-A feature that gives calling customers the option to change the caller ID status of their number on a per-call basis. When the calling directory number is private, the caller ID display will not be available to the called party.

Calling Party Pays (CPP)-A communication service that bills the calling party for the delivery of their call through a network (such as a mobile communication network or freephone service) or for the providing of information (such as a news service.)

CAMEL-See: Customized Applications For Mobile Enhanced Logic

Carrier Access Billing System (CABS)-A system that is used by network access providers to bill carriers for their customer's access to the network facilities.

Carrier Identification Code (CIC)-A 3-digit code that uniquely identifies a telecommunications carrier within the North American Numbering Plan (NANP). The CIC is indicated by an XXX in a Carrier Access Code where X can be any digit, 0 through 9. After an XXX code has been assigned to a carrier, the code is retained for use with either Feature Groups B (950A-0XXX, 950-1XXX) or D (10XXX) throughout the area served by the NANP. (See also: Carrier Access Code, pre-subscription, primary interexchange carrier.

CCS-See: Common Channel Signaling

CDR-See: Call Detail Record

Cell Switching-A term that describes how a mobile communication system transfers a communication path from one radio transmitter site (cell) to another cell as a mobile communication device (mobile telephone) moves throughout a cellular system.

Cellular Intercarrier Billing Exchange Roamer (CIBER)-A billing standard designed to promote inter-carrier roaming between cellular telephone systems. The CIBER format is developed and maintained by CiberNet. Cibernet is owned by the Cellular Telecommunications Industry Association (CTIA).

Centrex-Centrex is a service offered by a local telephone service provider that allows the customer to have features that are typically associated with a private branch exchange (PBX). These features include 3 or 4 digit dialing, intercom features, distinctive line ringing for inside and outside lines, voice mail waiting indication and others. Centrex services are provided by the central office switching facilities in the telephone network.

Centum (Hundred) Call Seconds (CCS)-A measurement of communication trunk usage that equals 100 seconds of continuous usage. Because the standard time interval for communication network engineering is based on activity over an hour, the system load is expressed in hundred call seconds (CCS) for one hour. A single hour has thirty-six hundred call seconds (equal to one erlang).

Channel Service Unit (CSU)-A CSU is the hardware that is used to assign communication channels from one or more data terminal equipment (DTE) devices to logical channels on a multi-channel communication circuit. The CSU can be customer premises equipment (CPE) or provided by the telecommunications service provider. The CSU is often combined with a data service unit (DSU).

CIBER-See: Cellular Intercarrier Billing Exchange Roamer

CIC-See: Carrier Identification Code

CIC-See: Circuit Identification Code

CIR-See: Committed Information Rate

Circuit Identification Code (CIC)-An information code that identifies a circuit between a pair of SS7 exchanges, for which signaling is being performed (14 bits in the ANSI version and 12 bits in the ITU version ISDN User Part).

Circuit Switch-A device or assembly that receives signals on input ports and provides a continuous connection (may be a physical or logical connection) to output ports.

Circuit Switching-A process of connecting two points in a communications network where the path (switching points) through the network remains fixed during the operation of a communications circuit. While a circuit switched connection is in operation, the capacity of the circuit remains constant regardless of the amount of content (e.g. voice or data signal) that is transferred during the circuit connection.

Circuit Validation Test (CVT)-A SS7 signaling procedure used to ensure that the communication path between two exchanges have sufficient and consistent translation data for placing a call on a specific circuit.

CLASS-See: Custom Local Area Signaling Services

Clearing House-A clearing house is a company or association that performs data and financial settlements functions between service providers. The clearing house receives data records from its member providers; after a validation process, the data is forwarded to the provider who will bill the customer. An additional role of the clearing house is to handle financial settlements between Service Providers. Clearing houses are particularly important for international roaming, particularly due to the need for conversion between different data record formats used by the various service providers, and the handling of currency exchange rates.

CLI-See: Calling Line Identification

C-Link-See: Cross Link

CLLI-See: Common Language Location Identification

Closed User Group (CUG)-(1- access restriction) A group of directory numbers sharing an access restriction such that any directory number can reach others in the group but cannot access outside numbers. (2- cellular system) Advanced features such as 4-digit dialing authorized for a closed group of users of the service. (3 - X25 protocol) In the X.25 packet-switching protocol , a facility indicating a virtual grouping of terminals that can communicate only with other members of that group. The feature can be extended to a closed user group with outgoing access, or a closed user group with incoming access.

Committed Information Rate (CIR)-Committed information rate (CIR) is a guaranteed minimum data transmission rate of service that will be available to the user through a network. Applications that use CIR services include voice and real time data applications. CIR can be measured in bits per second, burst size, and burst interval.

Some service providers allow users to transmit data above the CIR level. However, when data is transmitted above the CIR level, some of the data may be selectively discarded if the network becomes congested.

Common Channel Signaling (CCS)-A signaling technique in which signaling information relating to a multiplicity of circuits (trunks), is conveyed over a separate single channel by addressed messages. Common channel signaling system #7 ("SS7") is the primary system used for interconnection of telephone systems. SS7 sends packets of control information between switching systems.

The SS7 network is composed of its own data packet switches, and these switching facilities are called signal transfer points (STPs). In some cases, when advanced intelligent network services are provided, STPs may communicate with signal control points (SCPs) to process advanced telephone services. STPs are the telephone network switching point that route control messages to other switching points. SCPs are databases that allow messages to be processed as they pass through the network (such as calling card information or call forwarding information).

Common Language Location Identification (CLLI)-A standard code used by telecommunications systems to identify specific locations of a switching office or network element. A CLLI code is composed of 11 alphanumeric characters. The first four characters of the CLLI code are an abbreviated place name. Characters 5 and 6 are state abbreviations. Positions 7 and 8 identify a specific building, and 9,10, and 11 represent a particular piece of equipment.

Computer Telephony Integration (CTI)-CTI is the integration of computer processing systems with telephone technology. Computer telephony provides PBX functions along with advanced call processing and information access services. These services include, pre-paid telephony access control, interactive voice response (IVR), call center management, and private branch exchanges (PBX).

Connectionless Switching-A communications model in which stations can exchange data without first establishing a connection. In connectionless communications, each frame or packet is handled independently of all others.

Continuous DTMF-See: Continuous Dual Tone Multi-Frequency

Continuous Dual Tone Multi-Frequency (Continuous DTMF)-A process for some telephones (especially mobile telephones and PBX telephones) that allows a dual tone multifrequency (DTMF) signal to be sent as long as the key is depressed. Some telephones send DTMF for a predefined time limit (such as 100 msec). Short DTMF tones may not allow access to services such as voice mail and answering machines. However, long DTMF tones that are sent over a poor communication line (such as an analog radio channel) may be recognized as multiple digits if temporary lapses or distortion occurs during DTMF transmission.

Core Switch-A backbone switch that interconnects to other core switches and edge switches. Core switches maintain information about virtual paths that are connected through the network.

COT-See: Customer Originated Trace

CPE-See: Customer Premises Equipment

CPP-See: Calling Party Pays

Cross Link (C-Link)-A "C" (cross) link connects signal transfer points (STPs) that perform identical functions into a mated pair within an SS7 network. A "C" link is used only when an STP has no other route available to a destination signaling point due to link failure(s). Note that SCPs may also be deployed in pairs to improve reliability; unlike STPs, however, mated SCPs are not interconnected by signaling links.

CSU-See: Channel Service Unit

CTI-See: Computer Telephony Integration

CUG-See: Closed User Group

Custom Calling Services-Custom local area signaling services (CLASS) are telephone service features available in a local access and transport area (LATA) that are primarily based on information that can be processed inside the telephone network. CLASS features include call forwarding, caller identification, and three-way calling.

Custom Local Area Signaling Services (CLASS)-A set of telephone services and enhanced features available in a local access customers that may include calling number delivery or calling name delivery (CND), message waiting, and other features.

Customer Originated Trace (COT)-A Custom Local Area Signaling Services (CLASS) feature that allows a subscriber to initiate a call trace request message. This call trace feature temporarily stores the dialed digits and alerts the telephone service operator to "tag" the originator's number to allow authorities to investigate the originator of the unwanted or unauthorized call. Some of the call trace activation codes include *57 on a touchtone phone or the dialing of 1157 on a rotary (pulse) phone. If the call trace of the last call was completed successfully, an announcement should be heard. The service operator will usually release the call trace information to law enforcement agencies and a signed authorization from the subscriber may be required.

Customer Premises Equipment (CPE)-All telecommunications terminal equipment located on the customer's premises, including telephone sets, private branch exchanges (PBXs), data terminals, and customer-owned coin-operated telephones.

Customized Applications For Mobile Enhanced Logic (CAMEL)-Applications that operates on a "services creation node" in a GSM network. CAMEL allows a network operator to develop specialized services using advanced intelligent network (AIN) systems.

CVT-See: Circuit Validation Test

CW-See: Call Waiting

Data Link Layer-Layer 2 of the seven-layer Open Systems Interconnection (OSI) protocol model that facilitates the detection of and recovery from transmission errors. A data link connection is built on one or more physical connections.

Data Network-Data networks is a system that transfers data between network access points (nodes) through data switching, system control and interconnection transmission lines. Data networks are primarily designed to transfer data from one point to one or more points (multipoint). Data networks may be composed of a variety of communication systems including: circuit switches, leased lines and packet switching networks. There are predominately two types of data networks, broadcast and point-to-point.

Data Service Unit/Channel Service Unit (DSU/CSU)-Devices that combine the functionality of data service units (DSU) and channel service units (CSU) to adapt data from user communication systems to communication lines with multiple channels. The DSU portion as an interface between a customer's data terminal equipment and a data communication network. DSU are the digital equivalent of the analog modem and are translation codecs (COde and DECode) coupled with a network termination interface

(NTI). The CSU portion is used to coordinate communication from one or more data terminal equipment (DTE) devices to logical channels on a multi-channel communication circuit.

Data User Part (DUP)-The User Part in an SS7 message specified for data services.

Database-A collection of data that is interrelated that is stored in memory (disk, computer, or other data storage medium.)

Database Management System (DBMS)-A system that is controls access to, organization of, security, and application interfaces to information data.

DBMS-See: Database Management System

Demarc-See: Demarcation Point

Demarcation Point (Demarc)-The physical and electrical boundary between an end user's telecommunication equipment and the telecommunications network. The demarcation point establishes point of ownership and accountability.

Destination Point Code (DPC)-A part in the label of an SS7 signaling message which uniquely identifies, in a signaling network, the destination point of the message. In the form network-region-signaling point.

Diagonal Link (D-Link)-A "D" (diagonal) link connects a secondary (e.g., local or regional) signal transfer point (STP) pair to a primary (e.g., inter-network gateway) STP pair in a quad-link configuration in an SS7 network. Secondary STPs within the same network are connected via a quad of "D" links.

Dial Tone-A signal tone provided from a local telephone service provider that indicates that the telephone network is ready to send a call (to receive dialed digits). The dial tone signal is usually a combination of 350 Hz and 440 Hz signals.

Digital Service Unit (DSU)-A device that interconnects the customers digital telephone equipment to a telephone network.

Digital Signal Processor (DSP)-A digital signal processor (DSP) is an electronics device or assembly (typically an integrated circuit) that is designed to process signals through the use of embedded microprocessor instructions. The use of DSPs in communication circuits allows manufacturers to quickly and reliably develop advanced communications systems through the use of software programs. The software programs (often called modules) perform advanced signal processing functions that previously complex dedicated electronics circuits. Although manufacturers may develop their own software modules, DSP software modules are often developed by other companies that specialize in specific types of communication technologies. For example, a manufacturer may purchase a software module for echo canceling from one DSP software module developer and a modulator software module from a different DSP software module developer.

D-Link-See: Diagonal Link

DPC-See: Destination Point Code

DSP-See: Digital Signal Processor

DSU-See: Digital Service Unit

DSU/CSU-See: Data Service Unit/Channel Service Unit

DUP-See: Data User Part

EIR-See: Equipment Identity Register

Electronic Switching System (ESS)-A system that can connect incoming and outgoing digital lines together through the use of temporary memory locations. For an ESS system, a computer controls the assignment, storage, and retrieval of memory locations so that a portion of an incoming line (time slot) can be stored in temporary memory and retrieved for insertion to an outgoing line.

E-Link-See: Extended Link

End Office (EO)-The end office interconnects calls between local customers and the telephone network. Each end office switch can usually supply service up to 10,000 customers. In larger areas (such as a city), established LECs may have several EO switches. The EO switches are interconnected using a higher level tandem switch. If is a significant amount of calls regularly processed between end offices, they may be directly connected via high-speed communication lines (trunks).

Enhanced 800 Services-A call processing service that allows for the routing of 800 (or other toll free or freephone numbers) to be routed to different locations based on other criteria such as day, time of day, or caller location.

EO-See: End Office

Equipment Identity Register (EIR)-A database in a GSM telecommunications network that contains the identity of telecommunications devices (such as mobile phones) and the status of these devices in the network (such as authorized or not-authorized). The EIR is primarily used to identify mobile phones that may have been stolen or have questionable usage patterns that may indicate fraudulent use. The EIR has three types of lists; white, black and gray. The white list holds known good IMEIs. The black lists holds invalid (barred) IMEIs. The gray list holds IMEIs that may be suspect for fraud or are being tested for validation.

ESS-See: Electronic Switching System

Ethernet Frame Structure-The structure of data packet that is used in the Ethernet data network. The data packet frame length is variable and can go up to 1500 bytes. The fields within the data packet include a preamble for synchronization, 48 bit destination address, 48 bit source address, control fields, data packet, and error detection.

ETSI-See: European Telecommunications Standards Institute

European Telecommunications Standards Institute (ETSI)-An organization that assists with the standards-making process in Europe. They work with other international standards bodies, including the International Standards Organization (ISO), in coordinating like activities.

Extended Link (E-Link)-An "E" (extended) link connects an SSP to an alternate STP. "E" links provide an alternate signaling path if an SSP's "home" STP cannot be reached via an "A" link. "E" links are not usually provisioned unless the benefit of a marginally higher degree of reliability justifies the added expense.

FIB-See: Forward Indicator Bit

Fill In Signal Unit (FISU)-An SS7 signal unit message that only contains only error control and delimitation information. This is transmitted when there are no message signal units (MSUs) or link status signal units to be transmitted. The continuous reception of this packet for periods that no messages are sent indicates the link is operation and allows the quality level (error rate) to be monitored.

FISU-See: Fill In Signal Unit

F-Link-See: Fully Associated Link

Forward Indicator Bit (FIB)-A bit inside an SS7 signal unit that indicates the start of a retransmission cycle.

Fragmentation-Fragmentation is a technique that divides a data packet into smaller data packets so that they can be sent through a network that can only transfer small data packets. Fragmentation occurs during network transmission. When these packets are received at their destination, they are reassembled to their original data packet size.

Fully Associated Link (F-Link)-An "F" (fully associated) link connects two signaling end points (i.e., SSPs and SCPs) on an associated link within an SS7 network. "F" links are not usually used in networks with STPs. In networks without STPs, "F" links directly connect signaling points.

Gateway-A gateway is a communications device or assembly that transforms data that is received from one network into a format that can be used by a different network. A gateway usually has more intelligence (processing function) than a bridge as it can adjust the protocols and timing between two dissimilar computer systems or data networks. A gateway can also be a router when its key function is to switch data between network points.

Gateway Mobile Switching Center (GMSC)-A switching system that is used in a mobile communications network that also connects to other networks such as the public switched telephone network (PSTN).

Geographic Number Portability (GNP)-Geographic number portability involves the transfer of telephone numbers for telephone devices or services that are used outside the normal geographic boundaries of the service provider's original system or area. Geographic number portability allows a customer to keep their same area code when they move to new cities or other distant geographic regions.

Global Title (GT)-A routing name (such as customer-dialed digits) that does not contain explicit information to enable routing in a communication network. The global title is usually converted to an address that allows the network to setup or route information to its ultimate destination point.

Global Title Translation (GTT)-A process used in a common-channel signaling system (such as SS7) that uses a routing table to convert an address (usually a telephone number) into the actual destination address (forwarding telephone number) or into the address of a service control point (database) that contains the customer data needed to process a call.

GMSC-See: Gateway Mobile Switching Center

GNP-See: Geographic Number Portability

GR-303-A set of technical specifications that help define the next generation of digital loop carrier (DLC) interconnection.

GT-See: Global Title

GTT-See: Global Title Translation

H.245-A signaling control protocol that contains a library of transmission control messages for use in packet based multimedia communication systems. H.245 control signaling consists of the exchange of end-to-end H.245 messages between communicating H.323 endpoints. The H.245 control messages are carried over H.245 control channels. The H.245 control channel is the logical channel 0 and is permanently open, unlike the media channels. The messages carried include messages to exchange capabilities of terminals and to open and close logical channels.

H.323-H.323 is an umbrella recommendation from the International Telecommunications Union (ITU) that sets standards for multimedia communications over Local Area Networks (LANs) that may not provide a guaranteed Quality of Service (QoS). H.323 specifies techniques for compressing and transmitting real-time voice, video, and data between a pair of video-conferencing workstations. It also describes signaling protocols for managing audio and video streams, as well as procedures for breaking data into packets and synchronizing transmissions across communications channels.

H.450 Supplementary Services-Supplementary Services for H.323, namely Call Transfer and Call Diversion, have been defined by the H.450 series. H.450.1 defines the signaling protocol between H.323 endpoints for the control of supplementary services. H.450.2 defines Call Transfer and H.450.3 Call Diversion. Call Transfer allows a call established between endpoint A and endpoint B to be transformed into a new call between endpoint B and a third endpoint, endpoint C. Call Diversion provides the supplementary services Call Forwarding Unconditional, Call Forwarding Busy, Call Forwarding No Reply and Call Deflection. The "hooks" for Supplementary Services are specified in H.323 Version 2.

Head-A device that erases, records, or reads information from a hard disk or other media that is passed under it.

High-Speed Multimedia-High-speed multimedia usually refers to image based media such as pictures, animation or video clips. High-speed multimedia usually requires peak data transfer rates of 1 Mbps or more.

HLR-See: Home Location Register

Holding Time-The amount of time that a line connection or other shared resource is in use for a call or call attempt.

Home Location Register (HLR)-The part of a wireless network (typically cellular or PCS) that holds the subscription and other information about each subscriber authorized to use the wireless network.

Hotline-A restricted calling class that forces a telephone (usually a wireless telephone) to be connected to an operator regardless of the digits actually dialed. Hotline is typically used when a telephone is first sold or activated to allow activation after the customer has provided the information to register for service or when the customer has not paid their bill.

Hundred Call Seconds -See: Centum

IAM-See: Initial Address Message

IAP-See: Initial Alignment Procedure

IAP-See: Intercept Access Point

IAP-See: Internet Access Provider

IGFs-See: International Gateway Facilities

ILMI-See: Interim Link Management Interface

IMUX-See: Inverse Multiplexer

Inbound Call Center-A call center (group of customer service agents) that receives telephone calls from customers. Inbound call centers (teleservice centers) are often used in response to advertisements and direct marketing campaigns. The call routing to inbound call centers can be fixed (established telephone lines) or dynamically controlled (based on activity or skills based routing.)

Information Service-Information services involve the processing of information that is transferred through a communications system. Information services add value to information by generating, acquiring, storing, transforming, processing, retrieving, utilizing, or making available information via telecommunications. Examples of information services include fax store and forward, electronic publishing, text to voice conversion, and news services.

Initial Address Message (IAM)-A message in the SS7 system that is used in the ISDN User Part to initiate trunk signaling between two SSP's.

Initial Alignment Procedure (IAP)-A procedure that is used when a signaling link is activated for the first time or when a link connection is restored after a communication failure.

Integrated Service Unit (ISU)-The combination of a digital service unit (DSU) and channel service unit (CSU) into one device. The ISU interfaces a customer's digital telephone equipment to the formats used by a telephone network by adapting digital protocols, electrical levels and physical connections.

Integrated Services Control Point (ISCP)-A software system that integrates services control point (SCP) features that allow for the efficient deployment of customized intelligent network services.

Integrated Services Digital Network (ISDN)-A structured all digital telephone network system that was developed to replace (upgrade) existing analog telephone networks. The ISDN network supports for advanced telecommunications services and defined universal standard interfaces that are used in wireless and wired communications systems.

ISDN provides several communication channels to customers via local loop lines through a standardized digital transmission line. ISDN is provided in two interface formats: a basic rate (primarily for consumers) and high-speed rate (primarily for businesses). The basic rate interface (BRI) is 144 kbps and is divided into three digital channels called 2B + D. The primary rate interface (PRI) is 1.54 Mbps and is divided into 23B + D for North America and 2.048 Mbps and is divided into 30B + 2D for the rest of the world. The digital channels for the BRI are carried over a single, unshielded, twisted pair, copper wire and the PRI is normally carried on (2) twisted pairs of copper wire.

Integrated Signal Transfer Point (ISTP)-A signaling switch used in the SS7 common channel signaling network that integrates other functions (such as global title translation) with its signal transfer (message switching) functions. These transfer points are used to route signaling messages (packets) to other signaling transfer points or network parts.

Intelligent Peripheral (IP)-A type of hardware that can be programmed to perform a new intelligent network capability in an SS7 network. IP's perform processing services such as interactive voice response (IVR), selected digit capture, and feature selection and account management for prepaid services.

Intercept Access Point (IAP)-A place in a communication network where information and/or content is intercepted for the purpose of passing it to a law enforcement agency.

Interexchange Carrier (IXC)-Inter-exchange carriers (IXCs) interconnect local systems with each other. IXCs are also known as long distance carriers. In the US, from 1984 until 1997, IXC and LEC operating companies were legally required to refrain from engaging in directly competitive business operations with each other. Since 1997, one business entity can engage in both IXC and LEC business if it satisfies certain competitive legal rules. In Europe and throughout the rest of the world, the same PTT operators also usually provide inter-exchange service within their country. In any case, governments regulate how networks are allowed to interconnect to local and long distance networks.

For inter-exchange connection, networks as a rule connect to long distance networks through a separate toll center (tandem switch). In the United States, this toll center is called a point of presence (POP) connection.

Interim Link Management Interface (ILMI)-A interim specification that was developed by the ATM forum to allow network management functions between public networks, private networks, and end users. ILMI has some of the capabilities associated with simple network management protocol (SNMP.)

Interim Standard 124 (IS-124) Data Message Handler (DMH)-A standard billing communication protocol that allows for the real time transmission of billing records between different systems. IS-124 messaging is independent of underlying technology and can be sent on X.25 or SS7 signaling links. The development of the standard is primarily led by CiberNet, a division of the cellular telecommunications industry association (CTIA).

interLATA-Telecommunication services that cross from a local access and transport area (LATA) into another LATA.

International Gateway Facilities (IGFs)-Systems or equipment that provide access between telephone systems in different countries. International gateways may convert SS7 and other signaling formats between different signaling formats. These include ANSI standards, ITU standards, national variants of SS7 signaling standards, MF signaling, and R2 signaling. International gateways may also provide for transcoding services between mu-LAW PCM and A-LAW PCM speech coding.

International Signaling Point Code-The part in the label of a CCITT#7 or SS7 message that uniquely identifies each signaling point belonging to the international signaling network. It consists of a sub-field for the signaling area/network code (11-bit) and a sub-field that identifies a signaling point in a specific area or network (3-bit).

International Telecommunication Union (ITU)-A specialized agency of the United Nations established to maintain and extend international cooperation for the maintenance, development, and efficient use of telecommunications. The union does this through standards and recommended regulations, and through technical and telecommunications studies. Based in Geneva, Switzerland, the ITU is composed of two consultative committees: the International Radio Consultative Committee (CCIIR) and the Consultative Committee for International Telephony And Telegraphy (CCITT).

Internet Access Provider (IAP)-A company that provides an end user with data communication service that allows them to connect to the Internet. Internet access providers are also called Internet service providers (IAPs.)

Internet Protocol Telephony (IP Telephony)-IP telephone systems provide voice or multimedia communication services through the use Internet protocol (IP) networks. These IP networks initiate, process, and receive voice or multimedia communications using IP protocol. These IP systems may be public IP systems (e.g. the Internet), private data systems (e.g. LAN based), or a hybrid of public and private systems.

Internet Signaling Transport Protocol (ISTP)-A signaling protocol that used by PacketCable networks to provide SS7 type signaling capabilities.

Internet Telephony-Telephone systems and services that use the Internet to initiate, process and receive voice communications.

Internet Telephony Service Provider (ITSP)-An Internet Telephony Service Provider (ITSP) is a company or that supply telephone service using Internet Protocol telephony.

Intersystem Handover-A process where a mobile radio operating on cell site in one mobile system is reassigned to a new channel on a cell site in another mobile system. Intersystem handover requires signaling messages to be transmitted between the different systems.

Intersystem Signaling-Control signaling that occurs between systems.

Intersystem Signaling 41 (IS-41)-The application entity that dedicated to the communication aspects of intersystem signaling (such as in the SS7 system) that is used for the AMPS (analog), IS-136 TDMA, and IS-95 CDMA mobile communication systems.

Inverse Multiplexer (IMUX)-A device that divides a single telephone or data communication channel into two or more channels to be transported over multiple communication links. Inverse multiplexing may be in the form of frequency division (e.g. multiple radio channels on a coax line), time division (e.g. slots on a T1 or E1 line), or code division (coded channels that share the same frequency band) or combinations of these.

Inverse Multiplexing-The combining of information signals received on multiple communications channels to form a higher speed communication channel than is possible on a single independent communication channel. Inverse multiplexing has been used on wireless communication systems to allow high-speed digital video signals to be sent over cellular radio channels that have a limited maximum data transmission rate.

IP-See: Intelligent Peripheral

IP Telephony-See: Internet Protocol Telephony

IS-41-See: Intersystem Signaling 41

ISCP-See: Integrated Services Control Point

ISDN-See: Integrated Services Digital Network

ISDN User Part (ISUP)-The functional part of SS7 protocol that provides the call processing control signaling functions that are required to support basic communication services. Although is it based in the Integrated Services Digital Network (ISDN) signaling functions, ISUP is used for analog and digital call processing functions.

ISTP-See: Integrated Signal Transfer Point

ISTP-See: Internet Signaling Transport Protocol

ISU-See: Integrated Service Unit

ISUP-See: ISDN User Part

ITSP-See: Internet Telephony Service Provider

ITU-See: International Telecommunication Union

IXC-See: Interexchange Carrier

Key Telephone System (KTS)-Key telephone systems are (usually small) multi-line private telephone network that allows each key telephone station to select one of several telephone lines, place a line on hold, and call via an intercom circuit between key telephones. Key systems contain a central key service unit (KSU) that coordinates status lights and lines to key telephones ("Key Sets"). Early KTS system technology was based on electromechanical relay hardware. They required all the outside telephone lines to be connected to all of the key telephone sets in the installation. In addition, two additional pairs of wire were used in conjunction with each telephone line, one pair for the A/A1 connection indicating if that line is off hook at that particular key telephone set, and another pair to operate a small light to indicate the status of that line. Consequently, each key telephone set was connected to the central KSU via a thick cable containing 50 wires (25 pairs). Newer KTS systems typically use only 4 wires to connect the electronic KSU to each electronic key telephone set, and are often called "skinny wire" key systems. Modern electronic key systems are small microprocessor controlled switching systems and have some of the same advanced call processing features such as call hold, busy status, multi-line conference, abbreviated dialing, and station-to-station intercom that are available in a larger PBX.

KTS-See: Key Telephone System

Last Call Return-A telephony service that allows a telephone user to automatically call back the phone number of the last received incoming call. Last call return is normally accomplished by the customer entering the service code (e.g. "*69").

Layer Management Entity (LME)-An protocol layer interface used in the SS7 system that converts a protocol layer (e.g. MTP2) to a Layer

Management Interface (LMI) format that can communicate with a Management Information Base (MIB) associated with a device or system.

Layer Management Interface (LMI)-The interface between a management information base (MIB) and the protocol layers of a communication system, such as the SS7 system.

Layered Protocols-Protocols that are designed to communicate with higher or lower level protocols in a communication network. Each layered protocol performs a specific function and each layered protocol has specific ways to pass information to protocols that in layers directly above or below it.

LIDB-See: Line Information Database

Line Information Database (LIDB)-A DataStore application program that resides on a service control point (SCP) in the SS7 telephone signaling network and provides validation information for use in alternate billing services, such as telephone calling cards. LIDB data base contains up-to-date records of all working lines, including directory listing name, description of the type of dialing capability subscribed (rotary dial vs. touch-tone), calling card numbers, and other data required for validation services. The acronym LIDB often is pronounced "lid-bee."

Line Side Connection-Line side connections are an interconnection line between the customer's equipment and the last switch (end office) in the telephone network. The line side connection isolates the customer's equipment from network signaling requirements. Line side connections and are usually low capacity (one channel) lines.

Link Set-A set of signaling links in an SS7 network that connects a pair of adjacent nodes.

Link Status Signal Unit (LSSU)-A signal unit (data packet) that contains information about the status of the link (e.g. failure or errors) in which it is transmitted in an SS7 network. This packet is specific to the link and is not transmitted through the network.

LME-See: Layer Management Entity

LMI-See: Layer Management Interface

LNP-See: Local Number Portability

Local Number Portability (LNP)-LNP is the process that allows a subscriber to keep their telephone number when they change service provider in their same geographic area. Local number portability requires that carriers release their control of one of their assigned telephone numbers so customers can transfer to a competitive provider without having to change their telephone number. LNP also involves providing access to databases of telephone numbers to competing companies that allow them to determine the destination of telephone calls delivered to a local service area.

Location Routing Number (LRN)-A telephone number (e.g. 10 digit number) that is used to route calls to and end office switch that allows for the processing of portable (assignable) telephone numbers.

LRN-See: Location Routing Number

LSSU-See: Link Status Signal Unit

M3UA-See: MTP3-User Adaptation Layer

Management Information Base (MIB)-Management information bases (MIBs) are a collection of definitions, which define the properties of the managed object within the device to be managed. Every managed device keeps a database of values for each of the definitions written in the MIB. MIBs are used in conjunction with the simple network management protocol (SNMP) as well as RMON to manage networks. MIBs (referred to now as MIB-i) were originally defined in RFC1066.

Management Routing Verification Test (MRVT)-Testing that validates the routing tables in the SS7 network nodes. MRVT tests ensure that no routing loops or other routing anomalies are present.

MAP-See: Mobile Application Part

Media Gateway (MG)-A gateway that interfaces the PSTN to multimedia data communication systems as specified by MGCP. The media gateway is responsible interface the different types of media formats between the public and data networks.

Media Gateway Controller (MGC)-The media gateway controller is the portion of a PSTN gateway that acts as a surrogate call management system (CMS). The MGC controls the signaling gateway and the media gateway (MG). The protocols between the MGC and MG include media gateway control protocol (MGCP), MEGACO, and H.323. The MGC acts as a call agent coordinating sessions between devices. Signaling between MGCs (agents) may use SIP or H.323 protocol.

Media Gateway Controller Protocol (MGCP)-MGCP is a control protocol that uses text format messages to setup, manage, and terminate multimedia communication sessions in a centralized communications system. This differs from other multimedia control protocol systems (such as H.323 or SIP) that allow the end points in the network to control the communication session. MGCP is specified in RFC 2705 and it was first drafted in 1998. MGCP forms the basis of the PacketCable NCS protocol.

Mediation Device-A network device in a telecommunications network that receives, processes, reformats and sends information to other formats between network elements. Mediation devices are commonly used for billing and customer care systems as these devices can take non-standard proprietary information (such as proprietary digital call detail records) from switches and other network equipment and reformat them into messages billing systems can understand.

Message Signal Unit (MSU)-A signal unit (data packet) that carries the signaling information (messages) that are transmitted through an SS7 network. This MSU packet contains control flags (fields) that indicate the protocol that is being transmitted (e.g. mobile application part or ISDN user part) along with a variable length information (message content) field.

Message Transfer Part (MTP)-The functional part of a common channel signaling system which transfers signaling messages as required by all the users. The message transfer part also contains, for example, error control and signaling security.

MF-See: Multifrequency Signaling

MG-See: Media Gateway

MGC-See: Media Gateway Controller

MGCP-See: Media Gateway Controller Protocol

MIB-See: Management Information Base

Mobile Application Part (MAP)-A set of call processing messages, originally defined for use with GSM, for setup and control of wireless calls via the public switched telephone network. It is normally implemented in conjunction with SS7 call processing messages. The North American standard IS-41 is similar in principle but different in details.

Mobile Switching Center (MSC)-Switching system that are used for mobile communication networks (cellular, PCS, and 3G.) The MSC was formerly called the mobile telephone switching office (MTSO).

MRVT-See: Management Routing Verification Test

MSC-See: Mobile Switching Center

MSU-See: Message Signal Unit

MTP-See: Message Transfer Part

MTP3-User Adaptation Layer (M3UA)-M3UA MTP3-User Adaptation Layer is a protocol for supporting the transport of any SS7 MTP3-User signaling (e.g., ISUP and SCCP messages) over Internet Protocol (IP) using the services of the Stream Control Transmission Protocol. This protocol would be used between a Signaling Gateway (SG) and a Media Gateway Controller (MGC) or IP-resident Database.

Multifrequency Signaling (MF)-Multifrequency (MF) signaling is a type of in-band address signaling method that represents decimal digits and auxiliary signals by pairs of frequencies from the following group: 700, 900, 1100, 1300, 1500 and 1700 Hz. These audio frequencies are used to indicate telephone address digits, precedence, control signals, such as line-busy or trunk-busy signals, and other required signals.

Multiplex- (1-general) The use of a common channel to make two or more channels. This is accomplished either by splitting of the common-channel frequency band into narrower bands, each of which is used to constitute a distinct channel (frequency-division multiplex), by allotting this common channel to multiple users in turn, to constitute different intermittent channels (time-division multiplex), or by allowing the simultaneous transmission of channels using unique identification codes (code-division multiplex.) (2-frequency division) A multiplexing system in which different frequency bands are used by different channels, enabling many different channels to be carried by a single frequency bearer channel. (3-time division) A multiplexing system in which the original analog signals are converted into digital form. The digital signals (for each of many channels) are transmitted sequentially at different time instants. (3-code division) A multiplexing system in which the original signals are converted into digital form and multiplied by a unique identification code. The digital signals (for each of many channels) are transmitted in parallel using different code identifiers.

Multiplexer (MuX)-A device that conveys two or more telephone or data conversations or connections on a single channel or link. Multiplexing may be in the form of frequency division (e.g. multiple radio channels on a coax line), time division (e.g. slots on a T1 or E1 line), code division (coded channels that share the same frequency band) or combinations of these.

MuX-See: Multiplexer

NACD-See: Network Automatic Call Distribution

Network Automatic Call Distribution (NACD)-NACD is a is a call processing system that routes (distributes) incoming telephone calls to specific telephone sets or stations calls based on the characteristics of the call or network settings. These characteristics can include an routing on network congestion, time of day routing, and other criteria.

Network Gateway (NGW)-A media and signaling adapter (gateway) used in a network to interface between different types of networks. A network gateway can convert both the media and signaling control messages between the systems.

Network Layer-The Network layer performs the switching and routing of data through the network, controls the flow of data within the network, segments (divides) or reformats data packets if necessary between network types, and performs error control functions specific to the address decoding and routing functions. The network layer receives data for transmission from an upper layer (such as a transport or session layer) and converts it into network addressable data formats that can be transferred through a network or transmission line. An upper layer provides the network layer with the necessary addressing and network routing control requirements (e.g. priority codes) to allow the network layer to send data through the network. The location of the network layer within the protocol stack is usually above a physical layer and below a transmission or session layer. The network layer is layer 3 in the open system interconnection (OSI) protocol layer model.

NGW-See: Network Gateway

Nodal Multiplexer-A multiplexer that has the capability of dynamically routing channels onto different communication circuits.

Number Portability-Number portability involves the ability for a telephone number to be transferred between different service providers. This allows customers to change service providers without having to change telephone numbers. Number portability involves three key elements: local number portability, service portability and geographic portability.

The first part of the telephone number (NPA-NXX) usually identifies a specific geographic area and specific switch where the customer subscribes to telephone service. If a telephone number is assigned to another system (different NXX) in the same geographic area (same NPA), the interconnecting carriers (IXCs) connecting to that system must know which local system to route the calls based on the selected local service providers. In this case, the IXC must look up the local telephone number in a database (called a database dip) prior to delivering the call to the end customer.

Off-Hook-A electrical signal that occurs when a customer typically removes a telephone receiver off of its cradle, thus releasing the hook switch. When the hook switch is released (off-hook), this typically causes a drop in telephone line voltage due to connecting of the local loop telephone wires together. Automatic devices such as a computer modem can also initiate an off-hook signals.

OMAP-See: Operation, Maintenance and Administration Part

On-Hook-An electrical signal that occurs when a customer typically replaces a telephone receiver onto its cradle, thus opening the hook switch. When the hook switch is opened (on-hook), this typically causes a increase in telephone line voltage due to removal of the connection between the telephone wires on the local loop line.

OPC-See: Originating Point Code

Open Interface-A connection or access point between two assemblies or systems that is well defined and is readily available to manufacturers or users of the interface. Open interfaces are usually defined to encourage competition as multiple manufacturers can compete to produce products that have open interfaces.

Open Systems Interconnection (OSI)-A standard set of protocols developed by the International Standards Organization (ISO) and the CCITT to facilitate the open inter-connection of computers and data terminals to their applications, regardless of their type or manufacturer. The protocols specify seven layers: physical, link, network, transport, session, presentation, and application. Each layer performs specific functions for data exchange and is independent of the other layers.

Operation, Maintenance and Administration Part (OMAP)-The Application Entity that is dedicated to the communications aspects of the Operation, Administration and Maintenance of the Signaling System Network

Originating Point Code (OPC)-A identification code that is part of an signaling system (SS7) message that uniquely identifies the originating point of the SS7 message.

OSI-See: Open Systems Interconnection

Outbound Call Center-A call center (group of customer service agents) that originate telephone calls from customers. Outbound call centers (telemarketing centers) are often used by companies to solicit new business or to obtain statistical or other business related information.

Packet Segmentation-The dividing of a block or packet of data into several segments (pieces.) Packet segmentation is often performed to divide a large data packet into smaller data packets so that they can be sent through a network that can only transfer small data packets. When these packets are received at their destination, they are reassembled to their original data packet size. See fragmenting.

Packet Switching-A mode of data transmission in which messages are broken into increments, or packets, each of which can be routed separately from a source then reassembled in the proper order at the destination.

PC-See: Point Code

Permanent Virtual Circuit (PVC)-A PVC is a virtual circuit is manually created for a continuous communication connection.

After a permanent communications circuit is established, a data path (logical connection) is maintained.

PHY-See: Physical Layer

Physical Layer (PHY)-The physical layer performs the conversion of data to a physical medium (such as copper, radio, or optical) transmission and coordinates the transmission and reception of these physical signals. The physical layer receives data for transmission from an upper layer, such as the Open System Interconnection (OSI) Data Link layer, and converts it into physical format suitable for transmission through a network (such as bursts, slots, frames, and superframes). An upper layer provides the physical layer with the necessary data and control (e.g. maximum packet size) to allow conversion to a format suitable for transmission on a specific network type and transmission line. The physical layer is layer 1 in the OSI protocol layer model.

Point Code (PC)-A code that is assigned to each node (signaling point) in an SS7 network. Point codes can be unstructured 14 bit (ITU) point codes or structured 24 bits (ANSI) point codes. The structured ANSI point codes are divided (hierarchical structure) a network identification, network cluster, and network cluster element (the specific device or assembly). Point codes are transferred in the signaling messages that exchanged between signaling end points and they identify the destination and source of the signaling message. The assignment of point codes is managed by a government agency.

Point-To-Point Protocol (PPP)-A communications protocol that uses Transmission Control Protocol and Internet Protocol (TCP/IP) to allow end users (end points) to connect directly to the Internet via a communications connection (usually dialup connection). For Microsoft operating systems, this type of connection is managed by Dial-up networking (DUN).

Port- (1-general) A connection point between a computer or computer-based machine and other hardware devices. (2-network) A place of access to a device or network where energy may be supplied or withdrawn, or where the device or network variables may be measured. (3- software) The process of moving source code and executable programs from one computing system to another of a different type without substantive changes to the source code.

PPP-See: Point-To-Point Protocol

Prepaid Calling Card-A card that is issued by a telecommunications service provider that contains coded identification information that permits the card holder to initiate a call or request and receive an information service. Calling cards contain a number or code contained on a magnetic stripe that uniquely identify the card and authorized services to the system.

Presentation Layer-Layer 6 of the OSI model. This layer responds to service requests from the Application Layer and issues service requests to the Session Layer. The Presentation Layer relieves the Application Layer of concern regarding syntactical differences in data representation within the end-user systems. Note: an example of a presentation service would be the conversion of an EBCDIC-coded text file to an ASCII-Coded file.

PRI-See: Primary Rate Interface

Primary Rate Interface (PRI)-An standard high-speed data communications interface that is used in the ISDN system. This interface provides a standard data rates for T1 1.544 Mbps and E1 2.048 Mbps. The interface can be divided into combinations of 384 kbps (H) channels, 64 kbps (B) channels and includes at least one 64 kbps (D) control channel.

Prioritization-Frame and packet prioritization assigns different priority codes to packet that are transmitted through a communication network. This allows some frames or packets to receive a higher transmission priority for time sensitive data communications (such as packetized voice).

Private Line-A dedicated communications circuit that is leased by a customer from a telephone service provider for voice, data, or video services. While a private line may be connected through a switching facility, the connection resources are constantly dedicated to the customer who is leasing the line.

Protocol Adaptation-The process of adapting one protocol to another protocol. This may involve syntax changes (text format and command name changes,) timing relationships, and other functional processes.

Pulse Dialing-One or two types dialing that uses rotary pulses to generate the telephone number.

PVC-See: Permanent Virtual Circuit

Q.931-A telecom call processing signaling protocol that is used in telephone communication systems. The Q.931 protocol defines the messages and formats are control messages that are created by the end communication device. Some of the common information contained in Q.931 messages include call setup and tear down messages, called and calling party telephone numbers, and other access control signaling messages.

Quantization-The process of representing a value with a less precise value. In analog-to-digital conversion, a continuous analog value is represented by one of a finite number of quantized values. In lossy signal compression (such as an A-law encoding) one digital value is represented by another one which is usually not precisely the same. Except in lucky cases where the quantized value is exactly the same as the original, quantization introduces error (or noise).

Queuing-A process of delaying or sequencing messages. Queuing involves receiving requests for service, prioritizing these requests, storing them in appropriate order and transferring the messages when the facilities (channels) are available to send them.

Queuing systems may change the order of messages or services to be provided based on priority access. For example, communication requests from a public safety official may be given priority over a communication request from a consumer.

RADIUS-See: Remote Access Dial In User Server

Real Time Billing-Real time billing involves the authorizing, gathering, rating, and posting of account information either at the time of service request or within a short time afterwards (may be several minutes). Real time billing is primarily used for prepaid services such as calling cards or prepaid wireless.

Redundancy-A system design that includes additional equipment for the backup of key systems or components in the event of an equipment or system failure. While redundancy improves the overall reliability of a system, it also increases the number of equipment assemblies that are contained within a network. Redundancy usually increases cost.

Regenerative Repeater-A repeater that receives, amplifies, reshapes, and retransmits digital signals.

Remote Access Dial In User Server (RADIUS)-A network device that receives identification information from a potential user of a network service, authenticates the identity of the user, validates the authorization to use the requested service and creates event information for accounting purposes.

Remote Monitoring Specification (RMON)-Protocols that allow network management software to configure, poll and trap events on network elements.

Repeater- (1-general) A device or circuit that is located between transmitting and receiving devices to improve the quality the signal that is delivered between them. A repeater obtains some or all of the signal from the transmitter, amplifies and may adjust (change a frequency) or filter the signal, and retransmits the signal to the receiver(s). (2-LAN) In a local area network, a device which operates at layer 1 (physical layer) of the OSI reference model. This device does not inspect packets, but instead regenerates all input signals on its output(s). Repeaters were common in shared-media Ethernet based on IEEE 802.3 10-Base-2 and 10-Base-5 protocols. In recent years, the need for repeaters has been greatly dimished as new physical layer transmission technologies have provided better transmission capabilities.

Request For Comments (RFC)-A requirements or draft standard document created by a standards body that solicits comments from manufacturers, carriers and industry experts to finalize the standard. When used by the Internet engineering task force (IETF) every major Internet Protocol is specified first by an RFC. There are many RFC documents available and they are a significant method used to define Internet protocols and technical standards.

Reseller-A reseller buys network services in bulk from an existing carrier for resale to the public or other customers. The reseller provides sales and support services to the customer and the customer usually pays the reseller for the communication services it receives.

RFC-See: Request For Comments

RJ-45-A standard 8 wire modular connector. RJ-45 connectors are commonly used in telephone and data communication systems.

RMON-See: Remote Monitoring Specification

Rotary Dial-A process of dialing using a spring-loaded mechanical switch that produces pulses as it rotates through 10 positions (1 through 9, and 0). As the rotary dial turns, a switch briefly interrupts the loop current. The number of pulses per rotation is counted to determine the number dialed. A time pause between rotary dials is used to determine when additional digits are dialed or when the caller is finished dialing.

SCCP-See: Signaling Connection Control Part

SCCP User Adaptation (SUA)-SCCP User Adaptation layer is a protocol for the transport of any SS7 SCCP user Signaling (e.g. TCAP, RANAP or RNSAP messages) over IP Between two signaling endpoints.

SCE-See: Service Creation Environment

SCP-See: Service Control Point

SCTP-See: Stream Control Transmission Protocol

SCTP Packet-Stream control transmission protocol (SCTP) packets contain a common header and variable length blocks (chunks) of data. The SCTP packet structure is designed to offer the benefits of connection-oriented data flow (sequential) with the variable packet size and the use of Internet protocol (IP) addressing.

SDH-See: Synchronous Digital Hierarchy

SDLC-See: Synchronous Data Link Control

SDP-See: Session Description Protocol

SDP-Service Data Point

Selective Call Acceptance-A service feature that only delivers calls to their dialed destination if they are on a previously specified selective call acceptance telephone number list. Calls that are received by other numbers are provided with a pre-recorded announcement that states the number is not accepting their call or the call may be routed to an alternate directory number.

Selective Call Forwarding-A service feature that forwards calls to one (or multiple) telephone numbers dependent on the incoming call forwarding criteria. Selective call forwarding can be used to redirect calls of a specific type (such as fax calls) to a pre-designated number (such as an office fax machine.)

Selective Call Rejection-A service feature that restricts the delivery of calls to their dialed destination if they are on a previously specified call rejection telephone number list. Selective call rejection is used to block calls from undesired callers such as prank callers or harassing bill collectors. Calls that are received by numbers on the call rejection list are provided with a pre-recorded announcement that states the number is not accepting their call or the call may be routed to an alternate directory number.

Service Control Point (SCP)-The signaling control point (SCP) is a computer database that receives information request messages from the SS7 network and returns information that is necessary for the completion of calls or services. The SCP usually receives requests fro a service switching point (SSP) via signaling transfer points (STPs) that determine that additional information is necessary to complete the call (such as an 800 toll free/freephone destination number lookup).

Service Creation Environment (SCE)-A development toolkit that allows the creation of services for advanced intelligent network (AIN) that is used as part of the signaling system 7 (SS7) network.

Service Information Octet (SIO)-Eight bits contained in an SS7 message signal unit (MSU), comprising the service indicator and the sub-service field.

Service Level Agreement (SLA)-An agreement between a customer and a service provider that defines the services provided by the carrier and the performance requirements of the customer. The SLA usually includes fees and discounts for the services based on the actual performance level received by the customer.

Service Management System (SMS)-A computer system that administers service between service developers and signal control point databases in the SS7 network. The SMS system supports the development of intelligent database services. The system contains routing instructions and other call processing information.

Service Number Portability (SNP)-Service number portability allows a customer to take their telephone number to a different type of service provider. Service number portability involves determination of the type of service provider (e.g., wireless or wired) who is responsible for completing the call using the telephone number (e.g. area code and NXX.) Service number portability may differ from local number portability as the interconnection and call processing for different types of service providers may vary.

Service Switching Point (SSP)-In an Intelligent Network (IN), a stored-program controlled switching system that has the functional capability to differentiate intelligent network calls and interact with service control points (SCPs). SCP databases are accessed by the SSP in providing database query oriented services such as the 800 Data Base Service and Alternate Billing Services. (See also: Intelligent Network. SSP is an IN term for the Class 4/5 Switch that have SS7 capabilities. The SSP has an open interface to the IN for switching signaling, control and handoff.

Session Description Protocol (SDP)-A text based protocol that is used throughout to provide high-level definitions of connections and media streams. The SDP protocol is used with session initiated protocol (SIP). The

SDP protocol is used in the PacketCable system. SDP is defined in RFC 2327.

Session Initiated Protocol (SIP)-SIP is an application layer protocol that uses text format messages to setup, manage, and terminate multimedia communication sessions. SIP is a simplified version of the ITU H.323 packet multimedia system. SIP is defined in RFC 2543.

Session Layer-The session layer protocol coordinates the information transmission between endpoints during a communication session. The session layer receives requests for transmission from an application layer and converts it into network addressable data formats that can be transferred through a network or transmission line. The session layer usually establishes a communication session, coordinates the overall control of the session (such as handling retransmission and restart requests), and termination procedures. The location of the session layer within the protocol stacks varies dependent on the protocol. The session layer is layer 5 in the open system interconnection (OSI) protocol layer model.

SG-See: Signaling Gateway

Short Message Service (SMS)-A messaging service that typically transfers small amounts of text (several hundred characters). Short messaging services can be broadcast without acknowledgement (e.g. traffic reports) or sent point-to-point (paging or email). Most digital cellular systems have SMS services. Short messaging for mobile telephones may include: numeric pages (dialed in by a caller), messages that are entered by a live operator via keyboard, an automatic message service that sends a predefined message when an event occurs (such as a fire alarm or system equipment failure), network operator announcements to customers, to and from other message capable devices in the system, from the Internet, advertisers or other information providers.

Short Message Service Center (SMSC)-The facility that processes and routes short messages to telecommunications devices (such as a mobile telephone.)

SIF-See: Signaling Information Field

Signal Transfer Point (STP)-A signaling switch used in the SS7 common channel signaling network. These transfer points are used to route signaling messages (packets) to other signaling transfer points or network parts.

Signal Transport (SIGTRAN)-A set of standard that were defined by the Internet engineering task force (IETF) that contain a set of protocols that are suitable to provide signaling control messages (such as SS7 message) over in Internet Protocol (IP) network.

Signaling-The process of transferring control information such as address, call supervision, or other connection information between communication equipment and other equipment or systems.

Signaling Connection Control Part (SCCP)-The functional part of a common channel signaling system that provides additional functions to the MTP to cater for both connectionless as well as connection-oriented network service and to achieve an OSI compatible network service.

Signaling Gateway (SG)-A signaling gateway (SG) is used to interface a signaling control system (e.g. such as SS7) and a network device (e.g. a transfer point, database, or other type of signaling system). The signaling gateway may convert message formats, translate addresses, and allow different signaling protocols to interact.

Signaling Information Field (SIF)-The bits of a message signal unit (MSU) which carry information particular to a certain user transaction and always contain a label.

Signaling Link-A communication path that carries common channel signaling messages between two adjacent signaling nodes.

Signaling Message Discrimination-In the Signaling System 7 protocol, the process that decides if each incoming message, whether the signaling point is a destination point or if it should act as a signal transfer point (STP) for that message.

Signaling System 7 (SS7)-The signaling system #7 (SS7) is an international standard network signaling protocol that allows common channel (independent) signaling for call-establishment, billing, routing, and information-exchange between nodes in the public switched telephone network (PSTN). SS7 system protocols are optimized for telephone system control connections and they are only directly accessible to telephone network operators.

Common channel signaling (CCS) is a separate signaling system that separates content of telephone calls from the information used to set up the call (signaling information). When call-processing information is separated from the communication channel, it is called "out-of-band" signaling. This signaling method uses one of the channels on a multi-channel network for the control, accounting, and management of traffic on all of the channels of the network.

An SS7 network is composed of service switching points (SSPs), signaling transfer points (STPs), and service control points (SCPs). The SSP gathers the analog signaling information from the local line in the network (end point) and converts the information into an SS7 message. These messages are transferred into the SS7 network to STPs that transfer the packet closer to its destination. When special processing of the message is required (such as rerouting a call to a call forwarding number), the STP routes the message to a SCP. The SCP is a database that can use the incoming message to determine other numbers and features that are associated with this particular call.

In the SS7 protocol, an address, such as customer-dialed digits, does not contain explicit information to enable routing in a signaling network. It then will require the signaling connection control part (SCCP) translation function. This is a process in the SS7 system that uses a routing tables to convert an address (usually a telephone number) into the actual destination address (forwarding telephone number) or into the address of a service control point (database) that contains the customer data needed to process a call.

Intelligence in the network can be distributed to databases and information processing points throughout the network because the network uses common channel signaling A set of service development tools has been developed to allow companies to offer advanced intelligent network (AIN) services

Signal Transfer Point (STP)-A signaling switch used in the SS7 common channel signaling network. These transfer points are used to route signaling messages (packets) to other signaling transfer points or network parts.

Signaling Transport (SIGTRAN)-A set of standard that were defined by the Internet engineering task force (IETF) that contain a set of protocols that are suitable to provide signaling control messages (such as SS7 message) over in Internet Protocol (IP) network.

Signaling Unit Error Rate Monitor (SUERM)-An error rate monitor that is used in the SS7 system that is used to estimate the error rates associated with a signaling link.

SIGTRAN-See: Signal Transport

SIGTRAN-See: Signaling Transport

Simple Network Management Protocol (SNMP)-The Simple Network Management Protocol (SNMP) is a standard protocol used to communicate management information between the network management stations (NMS) and the agents (ex. routers, switches, network devices) in the network elements. By conforming to this protocol, equipment assemblies that are produced by different manufacturers can be managed by a single program. SNMP protocol is widely used via Internet protocol (IP) and operates over UDP well-known ports of 161 and 162. SNMP was originally defined in RFC1098 and is now obsolete and updated by RFC1157.

SIO-See: Service Information Octet

SIP-See: Session Initiated Protocol

SLA-See: Service Level Agreement

SMAE-See: System Management Application Entity

SMAP-See: System Management Application Process

SMS-See: Service Management System

SMS-See: Short Message Service

SMSC-See: Short Message Service Center

SNMP-See: Simple Network Management Protocol

SNP-See: Service Number Portability

Soft Switch-Soft switches are interconnection switching systems that can dynamically change its connection data rates and protocols types by software control to provide for voice, data, and video services. Soft switches were developed to replace existing end office (EO) switches that have limited interconnection capabilities. Soft switches are packet based and can simulate multiple protocols such as Internet protocol and asynchronous transfer mode (ATM). This allows for multiple types and simultaneous services to each customer with varying levels of quality of service (QoS).

SONET-See: Synchronous Optical Network

SS-See: Supplementary Service

SS7-See: Signaling System 7

SS7 Protocol Stack-The hardware and software functions of the SS7 protocol are divided into functional abstractions called "levels". These levels map loosely to the Open Systems Interconnect (OSI) 7-layer model defined by the International Standards Organization (ISO). Together these levels are called the SS7 Protocol Stack.

SSP-See: Service Switching Point

STP-See: Signal Transfer Point

STP-See: Signaling Transfer Point

Stream Control Transmission Protocol (SCTP)-A protocol that is used to coordinate the sending of signaling information over real time communication sessions.

SUA-See: SCCP User Adaptation

SUERM-See: Signaling Unit Error Rate Monitor

Supplementary Service (SS)-Supplementary services provide a network user with capabilities beyond those of elementary call control. Supplementary services enrich the basic services functions and are not specific to a telephone or system features. Often, the subscriber (user) can specify some of the operations of supplementary services (such as call forwarding). Supplementary services may be defined or installed in systems before complete testing or industry consensus can be reached.

SVC-See: Switched Virtual Circuit

Switched Virtual Circuit (SVC)-A switched virtual circuit is an automatically and temporarily created virtual connection that is used for a communication session.

Synchronous Data Link Control (SDLC)-A bit oriented synchronous communication protocol that organizes information into sequenced frames (groups of bits) of data. SDLC is similar to the high-level data link control (HDLC) protocol defined by the International Organization for Standardization (ISO).

Synchronous Digital Hierarchy (SDH)-A digital transmission format that is used in optical (fiber) networks to transport high-speed data signals. SDH uses standard data transfer rates and defined frame structures formats in a synchronous (sequential) format. SDH is similar to SONET.

Synchronous Network-A network in which the data communication lines are synchronized to each other or to a common clock signal that allows the exact determination of groups of bits (frames or fields) that are defined within the transmission of digital information.

Synchronous Optical Network (SONET)-A digital transmission format that is used in optical (fiber) networks to transport high-speed data signals. SONET uses standard data transfer rates and defined frame structures formats in a synchronous (sequential) format.

System Management Application Entity (SMAE)-An SS7 software function that provides communications functions that support applications that are part of system manage application process (SMAP). There may be several SMAE functions for each SMAP process.

System Management Application Process (SMAP)-A SS7 software function or process that monitors, controls, and coordinates resources for application layer protocols.

Tandem Free Operation (TFO)-Tandem free operation involves the direct connection of switching centers in mobile communication system without the need to decompress and re-compress (transcoding) speech information. TFO overcomes the challenges of cascading the speech coding process. Each time speech information is compressed and decompressed, some audio distortion occurs and time delay is added.

The logical function for coordinating TFO is the transcoder rate adaption unit (TRAU). The TRAU is usually located in the MSC (it is possible to put the TRAU in the base station) and it negotiates the ability of the MSC to use TFO with another MSC. The TRAU is also responsible for disabling TFO if the call is transferred to another MSC or system that is not capable of TFO.

TAP-See: Transferred Account Procedures

TCAP-See: Transaction Capabilities Application Part

TCP-See: Transmission Control Protocol

Telemarketing-Telemarketing is the process of conducting marketing and sales programs using telecommunication systems. Telemarketing call centers generally combine advanced call processing systems (e.g. computer telephony automatic call distribution) with customer order processing systems.

Telephone Company-Telephone companies (also known as service providers or carriers) provide communication services to the general public. They are usually are regulated by the government and in some countries, may be partly or wholly owned by the government. For most European countries and many other countries, local telephone service is provided by government owned posts, telephone and telegraph (PTT) operators. In some European countries, the post (mail) network has been separated from the operation of telephone and telegraph networks. In some countries, the telephone and telegraph systems have become privatized, and are no longer owned by the government.

Telephony-The use of electrical, optical, and/or radio signals to transmit sound to remote locations. Generally, the term means interactive communications over a distance. Often, telephony relates to a telecommunications infrastructure designed and built by private or government-operated telephone companies.

Temporary Location Directory Number (TLDN)-A temporary location directory number (TLDN) is a temporary identification number that is used to route calls from a home system and a visited communication system. TLDN numbers are commonly used in mobile communication systems where mobile telephones regularly operate in other (visited) systems. The TLDN is usually assigned for each call delivery request received from the home system. When the call is received into the visited system, it is mapped (translated) to the current resources (e.g. cell site and mobile number) that are currently being used in the visited system.

TFO-See: Tandem Free Operation

TLDN-See: Temporary Location Directory Number

Transaction Capabilities Application Part (TCAP)-The portion of the SS7 protocol which is responsible for information transfer between two or more nodes in the signaling network. The application layer of the Transaction Capabilities protocol consists of transaction capabilities that manage remote operations.

Transcoding-The conversion of digital signals from one coding format to another. An example of transcoding is the conversion of u-Law encoded PCM to A-Law encoded PCM signals.

Transferred Account Procedures (TAP)-Transferred accounting process (TAP) is a standard billing format that is primarily used for global system for mobile (GSM) cellular and personal communications systems (PCS). As of 2001, the versions of TAP, TAP II, TAP II+, NAIG TAP II, and TAP 3. Each successive version of TAP provided for enhanced features.

Due to the global nature of 3G wireless and GSM, the TAP billing standard provides solutions for multi-lingual and multiple exchange rate issues. TAP3 was released in 2000 as a significant revision of TAP2. TAP3 has changed from the fixed record size used in TAP2 to variable record size and TAP3 offers billing information for many new types of services such as billing for short messaging and other information services. The TAP standard is managed by the GSM association at www.GSMmobile.com.

Transmission Control Protocol (TCP)-A session layer protocol that coordinates the transmission, reception, and retransmission of packets in a data network to ensure reliable (confirmed) communication. The TCP protocol coordinates the division of data information into packets, adds sequence and flow control information to the packets, and coordinates the confirmation and retransmission of packets that are lost during a communication session. TCP utilizes Internet Protocol (IP) as the network layer protocol.

Transport Overhead-In the Synchronous Optical Network (SONET), the overhead added to the Synchronous Payload Envelope of a Synchronous Transport Signal for transport purposes. Transport overhead consists of line and section overhead.

Trunk Side Connection-Trunk side connections are used to interconnect telephone network switching systems to each other. Trunk side connections are usually high capacity lines.

U Law Encoding-The type of non-linear digital voice coding (digital signal companding) that is commonly used in the Americas and other parts of the world. The U Law (pronounced Mu Law) coding process is used to compress the 13 bit sampling of a digitized audio signal into the equivalent of an 8 bit sample. It does this by assigning a non-binary (non-linear) value to each of the binary bits. Another non-linear voice coding system is the A Law coding system that is used in Europe and other parts of the world.

UDP-See: User Datagram Protocol

UMTS-See: Universal Mobile Telecommunications System

Unbundling Services-Unbundling is the process of separating portions of a telecommunication network that are owned or operated by a service provider. Unbundling is a common term used to describe the separation of standard telephone equipment and services to allow competing telephone service providers to gain fair access to parts of incumbent telephone company systems. An example of an unbundled service is for the incumbent phone company to lease access to the copper wire line that connects an end user to the local telephone company. The competing company may install high-speed data modems (such as ADSL) on the copper line to enhancing the value of the telecommunications service.

Unified Messaging-A messaging application that provides storage, management, and delivery to different forms of messages. Unified messages include electronic mail (email), voice messages, and fax messages. Unified messaging involves the coordination of multiple access and control systems to control multimedia types of messages.

Unit Data-An SS7 message that contains information for ISuP and TCAP services.

Universal Mobile Telecommunications System (UMTS)-A Universal Mobile Telecommunications System (UMTS) that offers personal telecommunications services that uses the combination of wireless and fixed systems to provide seamless telecommunications services to its users. It is expected that UMTS will allow on-demand transmission capacities of up to 2 Mb/s in some of its radio locations. It should be compatible with broadband ISDN services.

User Datagram Protocol (UDP)-UDP is a high-level communication protocol that coordinates the one-way transmission of data in a packet data network. The UDP protocol coordinates the division of files or blocks of data information into packets and adds sequence information to the packets that are transmitted during a communication session using Internet protocol (IP) addressing. This allows the receiving end to receive and re-sequence the packets to recreate the original data file or block of data that was transmitted. UDP adds a small amount of overhead (control data) to each packet relative to other high-level protocols such as TCP. However, UDP does not provide any guarantees to data delivery through the network. UDP protocol is defined in request for comments 768 (RFC 768).

Value Added Services (VAS)-Services that provides benefits to a customer that are not part of the standard telecommunications services associated with a basic communication service. VAS services include voice mail, information services and content delivery.

Services offered by prepaid provider (e.g., voice mail, fax store and forward, interactive voice response, and information services) in addition to calling time.

VAS-See: Value Added Services

Virtual Private Network (VPN)-Secure private communication path(s) through one or more data network that is dedicated between two or more points. VPN connections allow data to safely and privately pass over public networks (such as the Internet). The data traveling between two points is encrypted for privacy.

Virtual Reality Modeling Language (VRML)-A text based language that is used to allow the creation of three-dimensional viewpoints, primarily for use with Web browsing. VRML was created by Mark Pesce and Tony Parisi in 1994 and is a subset of Silicon Graphics' Inventor File Format,

Visitor Location Register (VLR)-The database part of a wireless network (typically cellular or UMTS) that holds the subscription and other information about visiting subscribers that are authorized to use the wireless network.

VLR-See: Visitor Location Register

Voice Gateway-A voice gateway is a communications device or assembly that transforms audio that is received from a telephone device or telecommunications system (e.g. PBX) into a format that can be used by a different network. A voice gateway usually has more intelligence (processing function) than a bridge as it can select the voice compression coder and adjust the protocols and timing between two dissimilar computer systems or voice over data networks.

VPN-See: Virtual Private Network

VRML-See: Virtual Reality Modeling Language

X.25-An international standard for communications with a packet data switching network. The X.25 standard specifies the protocol between the data device (such as a computer) and the network (such as a public packet data network).

Index

Do you want to get more information or have an expert help you to understand how to use your data networks or the Internet to reduce your telecommunications costs?

Consider using Althos to educate you or your staff on the implementation and technologies used to connect telephones through data networks. Althos offers onsite courses, public courses, and real-time web seminar training. If you want instruction from experts who have setup Internet telephone systems, consider Althos. Althos can customize training to cover your key subject areas. Althos also has standard courses including:

Introduction to IP Telephony Systems

Basic IP Telephony Technology, How to Setup IPBX and IP Centrex, and IP Telephony Services & Economics.

How to Upgrade Web Sites for Mobile Devices

WAP, WML, cHTML, xHTML, Screens, Sounds, and Control, and Practice Code and Scripting

Introduction to Internet Billing

Real time and Post time Billing, Software and Hosting Options, Linking Information Systems Together, Record Formats, and Standards.

About Our Instructors

Mr. Harte is the president of Althos, an expert information provider covering the communications industry. He has over 29 years of technology analysis, development, implementation, and business management experience. Mr. Harte has worked for leading companies including Ericsson/General Electric, Audiovox/Toshiba and Westinghouse and consulted for hundreds of other companies. Mr. Harte continually researches, analyzes, and tests new communication technologies, applications, and services. He has authored over 30 books on telecommunications technologies on topics including Wireless Mobile, Data Communications, VoIP, Broadband, Prepaid Services, and Communications Billing. Mr. Harte holds many degrees and certificates include an Executive MBA from Wake Forest University (1995) and a BSET from the University of the State of New York, (1990).

Mr. Bowler is an independent telecommunications training consultant. He has almost 20 years experience in designing and delivering training in the areas of wireless networks and related technologies, including CDMA, TDMA, GSM and 3G systems. He also has expertise in Wireless Local Loop and microwave radio systems and has designed and delivered a range of training courses on SS7 and other network signaling protocols. Mr. Bowler has worked for a number of telecommunications operators including Cable and Wireless and Mercury Communications and also for Wray Castle a telecommunications training company where he was responsible for the design of training programmes for delivery on a global basis. Mr. Bowler was educated in the United Kingdom and holds a series of specialized maritime electronic engineering certificates.

Typical Training Costs

On-Site Training

Althos on-site training cost ranges from $2,200 to $3,600 per day of instruction plus expenses dependent on the length of the course and the type of content (labs, exercise materials, and instructor skill level). Althos does not charge for instructor travel time.

If Althos must incur travel expenses in conjunction with the project (this is typical for an on-site training session or presentation), travel expenses will be reimbursed on the basis of actual cost. The client prior to commitment will approve all travel expenses.

Online Training (Web Seminars)

Althos provides some courses and executive briefings in the form of online web seminars. Althos web training seminars allow two-way audio with all the participants along with presentation materials. The typical cost of web seminars range from about $85 for a 1-hour open enrollment executive briefing to approximately $350 per day for standard course instruction.

Open Enrollment

Althos periodically offers open enrollment to allow individuals to attend courses. The typical cost for individual enrollment ranges from $1,100 to $1,800 per student dependent on the location and type of course. Open enrollment courses include meals and materials (books and workbooks).

Custom Course Development

Althos can customize our courses to meet your specific training need or we can research and use your materials to create a new course. Custom course development fees range from $50 to $200 per presentation slide (graphics + descriptive text).

Printed in the United States
46534LVS00004B/8